Seismic Data Processing
THEORY AND PRACTICE

Seismic Data Processing

THEORY AND PRACTICE

L. HATTON MA, MSc, PhD, ALCN
Managing Director, Merlin Profilers (Research) Ltd

M. H. WORTHINGTON BSc, MSc, PhD
Professor of Geophysics, Imperial College of
Science and Technology, London University

J. MAKIN MA
Technical Director, Ensign Geophysics Ltd

BLACKWELL SCIENTIFIC PUBLICATIONS

OXFORD   LONDON   EDINBURGH

BOSTON   PALO ALTO   MELBOURNE

© 1986 by Blackwell Scientific Publications
Editorial offices:
Osney Mead, Oxford, OX2 0EL
8 John Street, London, WC1N 2ES
23 Ainslie Place, Edinburgh, EH3 6AJ
52 Beacon Street, Boston, Massachusetts 02108, USA
667 Lytton Avenue, Palo Alto, California 94301, USA
107 Barry Street, Carlton, Victoria 3053, Australia

First published 1986

DISTRIBUTORS

USA and Canada
Blackwell Scientific Publications Inc, P O Box 50009, Palo
Alto, California 94303

Australia
Blackwell Scientific Publications (Australia) Pty Ltd, 107
Barry Street, Carlton, Victoria 3053

British Library
Cataloguing in Publication Data

Hatton, L.
Seismic data processing: theory and practice.
1. Seismic waves—Data processing
I. Title   II. Worthington, M. H.   III. Makin, J.
551.2′2   QE538.5

ISBN 0-632-01374-5

Set by Santype Ltd, Salisbury, Wilts and printed and bound
in Great Britain by Butler & Tanner Ltd, Frome and London

Contents

Preface, vii

1 Seismic Data Processing and
   Computer Systems, 1
   1.1 Introduction
   1.2 The elements of a typical seismic processing
        system

2 Time Series Analysis in Seismology, 9
   2.1 Introduction
   2.2 Basic Fourier theory
   2.3 The discrete Fourier transform
   2.4 Convolution
   2.5 Filtering
   2.6 The Z-transform, phase and delay
   2.7 The auto- and cross-correlation functions
   2.8 The Weiner filter
   2.9 Spectral analysis
   2.10 The 2-D Fourier transform

3 Seismic Data Processing, 47
   3.1 Introduction
   3.2 Demultiplexing, gain recovery and tape formats
   3.3 Fundamentals of marine seismic data processing
   3.4 Data enhancement techniques
   3.5 Marine seismic data processing schemes
   3.6 Irregular and extended geometry
   3.7 Differences between land and marine processing

4 Seismic Migration, 123
   4.1 Migration and the three domains
   4.2 Natural migration coordinates
   4.3 Migration velocities and velocity sensitivity
   4.4 Migration and noise
   4.5 Migration and commutativity
   4.6 Migration and spatial aliassing
   4.7 Some useful migration formulae
   4.8 Migration before stack

5 Inverse Theory and Applications, 139
   5.1 Introduction
   5.2 Non-uniqueness
   5.3 Non-linear methods
   5.4 Matrix formulation of the inverse problem
   5.5 Linearisation
   5.6 Generalised matrix inversion
   5.7 Practical problems with GMI
   5.8 Damped least squares
   5.9 VSP processing
   5.10 Residual statics
   5.11 Linear programming
   5.12 Tomography
   5.13 Appendix

Problems, 165

References, 169

Index, 171

Preface

Most of the contents of this book originated as lecture notes for the M.Sc. course in Petroleum Exploration Seismology at Oxford University, and subsequently the M.Sc. course in Exploration Geophysics at Imperial College, London University. Students on these courses are typically drawn from a variety of backgrounds: recent graduates in geology, physics, mathematics, engineering, geophysics or some other combination of the natural sciences, and employees from a wide range of companies associated with hydrocarbon or mineral exploration. The mathematical competence of the participants has always been variable. Hence the mathematical content of this book has been kept to a minimum. However, we do assume that the reader understands the basic principles of differential and integral calculus, Fourier Series and simple matrix algebra, including the concepts of eigenvectors and eigenvalues and diagonalisation.

We are very conscious of other excellent books that have covered some of the same topics that we have discussed, and in many cases have done so with considerably more rigour. Our motivation for writing this book came from the belief that practical advice and instruction was in short supply. We assume that the reader has access to a computer system which is well equipped with relevant software, and the question is how this should be employed to its best advantage.

Chapter 1 provides a brief introduction to the computer hardware normally used in the seismic data processing industry. Chapter 2 provides the theoretical basis for the more practical Chapter 3. However, note the relative absence of integral signs compared to summations. Since, in practice, we accomplish everything with digital computers our emphasis is firmly with the discrete Fourier Transform. In Chapters 3 and 4, solutions to problems that must be faced countless times a day in any seismic data processing laboratory are described. We are indebted to Merlin Profilers Ltd and Ensign Geophysics Ltd for the use of their computer systems, software and graphics facilities, and to ARCO and Merlin Profilers Ltd for permission to publish the data used in these chapters. Chapter 5 is an attempt to present geophysical inverse theory in as digestable a form as possible. Most mini- and mainframe computer systems will have within their subroutine library all the code necessary to perform any operation described in this chapter, with the possible exception of some of the tomography algorithms.

Many students and colleagues have helped us with this book either directly or by exerting some less quantifiable influence. We are most grateful to them all. However, we would specifically like to thank Alec Gorski for his help with much of the discussion of f–k techniques in Chapters 2 and 3; Stuart Smith for his contributions to Section 3.6; Gregg Parkes for his contributions to Chapter 4; Mike Oristaglio, Shane Stainsby, Mark Loveridge and Iain Mason for some ideas incorporated in Chapter 5; Paddy Bennett and Andy Wright for proof reading and helpful comments; Ensign Geophysics Ltd for allowing John Makin to complete his contribution; and our respective families for putting up with the inevitable aggravation that our literary ambitions have provoked.

July
1985

Leslie Hatton
Michael Worthington
John Makin

Chapter 1

Seismic Data Processing and Computer Systems

1.1 Introduction

Since the mid 1960s, all seismic data processing has been implemented on a digital computer. Indeed, exploration seismology has been one of the important catalysts in the development of certain types of 'number-crunching', to use a hoary old aphorism. It is our intention in this chapter to discuss those practical aspects of computer systems which most affect the seismologist in the day-to-day practice of seismic data processing. This chapter acts as no more than an introduction to the jargon and a brief description of the equipment used. For a more complete description, the various papers by Hatton (1983a, b, c, d; 1984a, b) may be consulted. The material in this chapter is loosely based on the first of these papers.

It is common to find the mechanics of seismic data processing considered as a 'black art', amongst both geophysical programmers and the geophysical users of their products. One of the main reasons for this is that seismic data processing has traditionally placed such a heavy load on the computer technology of the time. At any stage in the last twenty years there have always been a number of processes which could be applied to the data but were not economically feasible. Nothing has changed. Algorithmically, full three-dimensional depth migration before stack could be implemented today if the velocity field were sufficiently accurately determined, but it would take the world's most powerful computers many years to do it. In practice, of course, the velocity field is, at best, a rather feeble approximation, but discussion of this vital problem will be temporarily postponed. Even something as mundane as demultiplexing can induce early retirement in a computer centre manager if the tapes concerned happen to be uncorrelated 30 second sweep 2 msec Vibroseis* at 6250 fpi. See later for an explanation of this terminology.

Such basic difficulties in implementing geophysical techniques have led to the establishment of the geophysical computing 'guru' who, by dint of erudite and obscure solutions, manages to circumvent the problems. Regrettably, such solutions tend to be extremely difficult either to maintain (correct errors), or enhance (add new functionality), due both to their complication and their dependence on the particular computer hardware on which they were developed. Consequently, users and programmers are unable to make easy use of improvements in computer software and hardware as they

become available, and the whole system rapidly achieves stasis.

Another important reason why the whole subject of the computer processing of seismic data is so labyrinthine is the quality of the documentation. The average piece of computer documentation is so appalling, not because it expresses deep mysteries, but because it is appallingly written. There are certain notable exceptions, but regrettably few of them. Unfortunately, it is not only the computer industry which produces poor documentation. The whole question of how the normal geophysical user 'interfaces to' or uses a particular seismic data processing system is often devalued relative to the importance of the sheer mechanics of 'getting the data through'.

Five years ago or so, computers were subdivided into three sizes, commonly known as microcomputers (the least powerful), minicomputers, and mainframes (the most powerful). A more accurate description could be given in terms of the intelligence of the peripheral devices attached to each category (discs, tapes, printers, etc.), in that microcomputer peripherals were the least intelligent (sophisticated), and mainframe peripherals the most intelligent. At the present rate of progress, the only difference will be in price in the not too distant future. At the time of writing, several microcomputers have been benchmarked as exceeding the power of a VAX 11/780, a well-known and deservedly respected minicomputer of considerable power.

1.2 The elements of a typical seismic processing system

The basic elements of a typical medium- to large-scale computer system for seismic work are shown in Fig. 1.1. The individual components will now be described in more detail.

1.2.1 The central processing unit

The central processing unit, normally abbreviated to CPU, governs the running of the whole computer system. Programs are fetched piece by piece from memory and are executed in the CPU, causing the various peripheral devices to respond in their own particular ways. The power of a CPU is commonly measured in a unit called a MIP, which is short for a million instructions per second. Unfortunately, this unit is rather ill-defined as it is extremely reliant on the power and flexibility of a particular CPU's instruction

---

*Trade mark of Conoco.

1

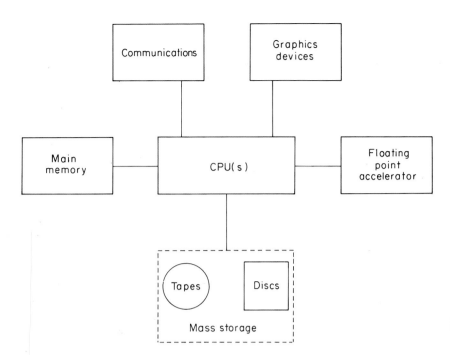

Fig. 1.1. Elements of a typical seismic data processing system.

set. A standard set of benchmark programs known as the Whetstone suite exists in order to provide a common denominator known as the Whetstone rating. Even then, many other factors influence a computer's performance in a given environment, especially operating system software, and two machines with nominally the same MIP or Whetstone ratings, may perform very differently in the seismic environment. Note that MIPs and mega-Whetstones are often confused. When quoting a MIP rating, a computer manufacturer will normally choose a fast instruction and compute how many times this particular instruction can be executed in a second. The Whetstone suite particularly references floating point operations which are not amongst the fastest of instructions, and thus the mega-Whetstone rating will usually be less than the MIP rating.

Many different types of computer are used for seismic data processing, ranging from the newest developments in microprocessor technology using chips manufactured by companies like Intel, Motorola, National Semiconductor, Inmos, Zilog, Hitachi, NEC and many others, all the way up to the giant computers or 'supercomputers' like the Cray and the CDC Cyber 205. Membership of this latter, somewhat exclusive club is restricted to computers which are capable of sustaining 20 megaflops of vector arithmetic performance. This concept will be explained shortly.

Microprocessor technology in particular, will have a very big part to play in the development of seismic computing systems, simply because of the enormous impact they are already having on the computer industry in general. During the period 1983–1986, microprocessors like the National Semiconductor NS32032 and 32132, Motorola MC68020, Intel 80286, 80386 and 80486, Inmos T424 'transputer' and Zilog 80000, along with a full range of 'support chips' including very powerful floating point processors and channels, are already threatening or will soon threaten the traditional large minicomputers and even mainframes in terms of seismic data processing capabilities. The price/performance of such chip sets is already staggering, with no apparent signs of a lessening of development activity. The approximate relative CPU performance of some microcomputer chips compared with various minicomputers, mainframe and supercomputers encountered in seismic data processing is shown in Table 1.2.1.

Table 1.2.1. The approximate relative CPU performance of various computers in use in the seismic data processing industry compared with microchip performance. The figures take no account of software and cannot be directly equated with seismic data throughput capacity.

Computer	Type	Performance
NS32132	Micro	2–3
MC68020	Micro	2–3
80386	Micro	2–3
Z80000	Micro	2–3
T424	Micro	8
VAX 11/780	Mini	1
Data General MV10000	Mini	2–3
Perkins-Elmer 3250	Mini	2–3
Norsk Data ND 570	Mini	3–4
IBM 3081	Mainframe	20
UNIVAC 1100/84	Mainframe	7
Cray 1S	Super	50–100
Cray X/MP	Super	100–200
CDC Cyber 205	Super	100–200

Main memory

In this section, the main memory will also be mentioned, without which the CPU is totally useless. Main memory consists of MOS (metal oxide semiconductor) technology electronic components nowadays, organised either into bytes or words.

Byte organisation A byte consists of eight bits (a bit is represented by an electronic device known as a flip-flop which is either in a high or a low voltage level (typically 5 and 0 volts respectively)). This electronic duality forces the use of binary arithmetic and its extensions (for example, octal (base 8) and hexadecimal (base 16)) when dealing with computers. The size of a computer's main memory is measured in Mb, or megabytes. A megabyte is normally understood as 2 to the power 20 bytes (a binary million, or 1024 times 1024). A byte-addressable machine is one in which the smallest individual addressable memory unit is a byte. Most micro- and minicomputers are byte-addressable.

Word organisation A computer word is not a well-defined concept in that there is little standardisation on word length. On micro- and minicomputers a word is usually two or four bytes although, as mentioned above, the memory itself is byte-addressable. On mainframes, a word is almost invariably four bytes, although there are some oddities like a 36-bit word. On all current supercomputers, the word length is eight bytes. On both mainframes and supercomputers, the memory is normally word-addressable, although it should be noted that almost all of the computers in Table 1.2.1 transfer information from and to main memory in four or eight byte words whether the memory is byte- or word-addressable. These transfers are done down highways known as buses which contain 32 or 64 parallel wires for four or eight byte data respectively.

It is commonly recognised that any kind of successful scientific computing requires a word length of at least four bytes. This is to prevent the growth of round-off errors during extended numerical calculations. In exploration seismic data processing, four byte words are entirely adequate, and it is only in certain fields, such as reservoir engineering, where eight bytes are considered necessary. Four bytes are certainly sufficient in any environment where noise of one kind or another contaminates the data.

Cache memory

A fairly large machine will have typically between 2 and 32 Mb of main memory separated from the CPU, usually by a much smaller amount of very high speed memory (called cache memory). The computer hardware uses this memory to hold what it thinks the next few instructions and data from the user's program will be, so that it can get them faster. If it guesses incorrectly (as it will if the user program is large and usage of each part is somewhat uniform, which is common in modern seismic data processing packages), then the effectiveness of cache memory is greatly diminished. Memory is getting very cheap nowadays and memory sizes will continue to rise. Ten years ago, memory was probably the most expensive part of a computer system, and consequently much smaller main memories were used (64–128 Kb (kilobytes)). Today, the cache memory alone is often this size!

1.2.2 Tape drives and tapes

Tapes are still by far the most common medium for the storage of very large amounts of data, hence their popularity in the seismic industry in which a typical 3-D seismic survey might form a dataset of some 100 000 megabytes (considerably more than 1000 tapes). The most popular tape technology is built around the 9-track tape, in which there are 9 recording tracks along the tape, of which 8 are used for data (for example a byte of information) and one is used for parity. The parity bit is written by tape hardware as an error-checking device. It is data dependent in the sense that if a byte is to be written to (i.e. recorded on) tape, the electronics controlling the tape drive on which the tape is mounted (a tape controller) check to see how many of the data bits are set to 1 (as opposed to 0), and the parity bit is set to ensure that the total number of 1's including the parity bit is odd (odd parity) or even (even parity). Odd parity is the most common. If the tape controller, when reading an odd parity tape, comes across a frame (all 9 tracks at any point on the tape) in which the number of bits set is even, an error is reported. Errors are fairly common on magnetic tape. For example, small pieces of dust can adhere to the tape. In this case, the tape controller will automatically rewind the tape to a position before which the error was reported and re-read. This may occur several times before the foreign matter is dislodged. If the foreign matter is peanut butter or mud, as can happen on tapes originally recorded by seismic field crews, this is insufficient, and the tape may have to be cleaned. In addition, some tapes have variations in the amount of magnetic material on them causing irrecoverable errors, but this is becoming increasingly rare at the present time.

In the seismic industry, tapes occur at three different recording densities, 800, 1600, and 6250 fpi, representing three different recording techniques, known as NRZI, PE and GCR recording. The fpi stands for frames per inch. So a GCR recording involves recording 6250 frames per inch, i.e. 6250 bytes per inch. Cpi (characters per inch) and bpi (bits per inch) are often used with the same meaning as fpi, but fpi is most accurate and preferable. Paradoxically, the highest density is also the most reliable. As many experienced seismologists will remember, 800 fpi NRZI tapes were notorious for the skew problems which occurred whenever the tape was not quite orthogonal to the reading heads on the tape drive. Physically, data appears on a magnetic tape as shown in Fig. 1.2.

The IRG, or inter-record gap, is a short piece of erased tape which can be sensed by the tape drive. Its length is 0.5 inch in 800 and 1600 fpi tape formats and 0.3 inch in the 6250 fpi tape format. In between IRGs, the data is called (by definition) a record. Unfortunately, this word is used indiscriminately for many different concepts in both geophysics and computer science, so it is best preceded by the qualifying word 'tape'. Although different computer systems handle tapes in different ways, from the point of view of the high-level

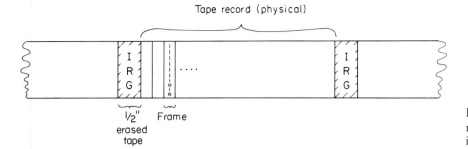

Fig. 1.2. Schematic of a physical tape record. IRG is an abbreviation of inter-record gap.

programmer and user (someone who treats the computer largely as a black box), the tape hardware reads everything between adjacent IRGs and cannot be interrupted in the middle. During this procedure, the whole tape record is read into the main memory. This causes many problems with seismic field tapes because whereas typical tape records may contain a few thousand bytes of information at the most, a seismic field tape record in one of the industry standard formats SEG A, B, C or D, will generally contain a million bytes or more. This process requires concurrent writing to disc as the tape spins owing to the fact that the amount of main memory available is usually insufficient. An analogy is baling out a leaking boat. If the ladle (disc) is insufficient to keep up with the hole (tape), the boat sinks (the tape is generally rewound and tried again).

It is worthwhile to illustrate some of the above points with a few simple calculations.

(a) Efficiency of tape usage

Consider the most common standard for demultiplexed seismic data, SEG Y. In this standard, a single physical tape record contains all the numbers pertaining to a particular seismic trace in IBM floating point format. This format requires 4 bytes or 32 bits per number. Suppose the seismic trace consists of 6 seconds of 4 msec data or 1500 samples. Then the physical length of the equivalent tape record on a 6250 fpi tape is

$$\frac{1500 * 4 \quad \text{(bytes per trace)}}{6250 \quad \text{(bytes per inch)}}$$

$$= 0.96 \text{ inch.}$$

Hence in this format with an IRG of 0.3 inch,

$$\frac{0.3}{(0.96 + 0.3)}$$

$$= 23.81\% \text{ of the tape is blank!}$$

(b) Storage capacity of a tape

The vast majority of field tapes acquired in recent years are 2400 feet tapes recorded at 1600 fpi. Excluding the headers which have a very small effect, for a SEG B format field tape with 120 channels of 6 second data at 2 msec, the length of a single physical tape record is

120	(channels)
* 3000	(samples)
* 2.5	(bytes per sample in SEG B)
= 900 kilobytes.	

This will use up

$\dfrac{900\,000}{1600}$	(length in bytes) (density in bytes/in)

$$= 562.5 \text{ inches}$$

$$+ 0.5 \quad \text{(IRG at 1600 fpi)}$$

$$= 563 \text{ inches per record.}$$

Hence a field tape can record up to

$$\frac{2400 * 12}{563} \quad \text{(length in inches)}$$

$$= 51 \text{ complete shot records.}$$

For a 'pop' rate of 40 shots per km, this gives only about 1.25 km per tape!

Recording and processing

The current state of the art in acquiring marine seismic data is to record 240 channels at 2 msec onto 6250 fpi tapes in a demultiplexed format like SEG Y. Repeating the above calculations, at 40 shots per km with a 6 second recording, a new tape will be required about every

$$\frac{2400 * 12}{2400 * \left(\dfrac{3000 * 4}{6250} + 0.3 \right) * 40}$$

$$= 1.35 \text{ km.}$$

Comparing this with the older technology equivalent in storage capacity above, illustrates another symptom of the classic problem which has always faced exploration seismologists and which has already been alluded to earlier. Here the major technology change of recording demultiplexed full floating point data at the tape density of 6250 fpi is almost exactly balanced by an increase in the number of channels and samples leaving essentially the same tape burden as before.

In the processing centre, the problem of inefficient tape usage for SEG Y tapes recorded at 6250 fpi can be overcome by using record blocking as is illustrated in Fig. 1.3. When the tape hardware reads the tape, it still transfers everything between IRGs, but a separate program in the computer is used to split up the composite or blocked record into its components. Unfortunately, on almost all computers in use in the seismic industry, this technique is geared for very short tape records rather than the long or very long records nor-

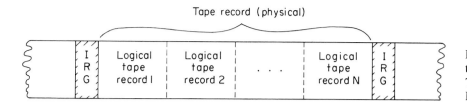

Fig. 1.3. Schematic of a blocked physical tape record, comprising N logical records. There is no physical distinction (such as an IRG) in between adjacent logical records.

mally encountered. Hence it would be a big step forward in seismic computing if computer manufacturers could include large amounts of buffer memory, for example a megabyte or more, in the tape controllers to take this load off the CPU. A considerable increase in throughput would certainly result. This point is particularly well appreciated in the microcomputer world, where even the latest printers have large internal memories and microprocessors to achieve such ends.

Two other identifiers are recognised by tape hardware to separate recordings.

The EOF

This is short for the end of file. This is a special character written on tape and invisible to the high-level user in that the character is never passed to the user, instead end of file status is returned, via the operating system. An EOF occurs between every shot record on seismic field tapes.

The EOM

This is short for the end of medium. It is a small piece of silver foil physically attached to the tape and intended to stop the single tape reel from unwinding off its spool. When the tape hardware senses the presence of this foil, it returns an end of medium status to the operating system, whence it is passed on to the user.

Sequential recording and access-time

Note finally that tape is a sequential medium. This means that, practically, it can only be read through record by record, in the order in which they appear on the tape. Consequently the time taken to access a particular record, known as the access-time, can be long. On the other hand, far more data can be recorded on tape than any other medium in current usage. This illustrates the basic trade-off in all computer systems of speed of access versus quantity of data to access.

1.2.3 Discs

The magnetic disc is a much faster storage medium than magnetic tape but can store much less information. Even so, this can be a fairly formidable amount. Currently, disc drives are available which can store anywhere between 20 and 600 Mb. The largest would be capable of storing the whole of *Encyclopaedia Britannica* and of accessing any individual word within 20 milliseconds—ideal for the quick reader. Figs. 1.4 and

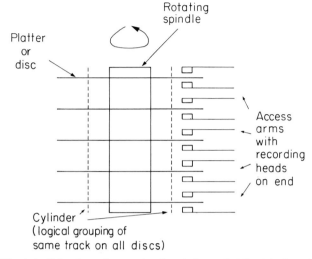

Fig. 1.4. Side view of a moving-head disc unit. The whole unit revolves at a speed usually of 3600 revolutions per minute.

1.5 show a side view and a top view respectively of a typical moving-head disc drive. The disc assembly rotates inside the disc drive normally at 3600 rpm. The disc assembly itself consists of a number of magnetic discs or platters all placed on the same spindle. Over these discs, small magnetic recording heads dart in and out driven by motors. The heads are very close indeed to the surface of the disc. An analogy commonly used is that of flying a jet at 600 mph a few inches off the ground. On occasions, discs fail catastrophically in what

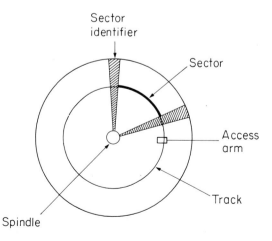

Fig. 1.5. Plan view of a moving-head disc unit. The sector identifiers are 'soft' (as IRGs are on tapes) and are written onto the disc when the disc is initialised. This process is known as formatting and is normally done once only. Until it is done, the disc cannot be used. The position of these identifiers is dependent on the computer system using them. Hence discs formatted on one system will usually not work on a different system.

is quaintly called a head crash. This occurs when the heads hit the rotating disc and are literally ripped from their mount, wrecking the disc surface at the same time. This may occur in power failure or if the platters become slightly misaligned. It is equivalent to having a piston come through the side of a car's engine in terms of the engineering effort required to correct it.

Other errors can occur with discs, just as they do with tapes. For example, a common type of error with disc drives is mis-positioning of the heads over the incorrect track. This occurs infrequently on modern discs, even though the tracks are very close together. It is normally corrected by returning the head to some reference position where its position can be calibrated before re-issuing the seek for the desired track.

In order of increasing storage, the disc is commonly split by a mixture of fixed and variable factors into:

> Sectors of 20–500 words.
> Tracks of 2000–5000 words.
> Cylinders of 40 000 words upwards.

Note that tracks have the same number of words independently of their distance from the spindle.

Disc access-time

In order to retrieve a piece of data from disc, the disc hardware must do the following. The only thing the user sees is the accumulated time delay.

1 Seek. In general, the recording heads will not physically be above the correct track. Hence, on average, they must move half-way across the disc before being correctly positioned. For most disc drives, this takes between 15 and 50 msec with the modern Winchester technology representing the lower end.

2 Latency. Since the disc is rotating, the recording heads must wait on average for half a revolution before the start of the data to be accessed lies underneath the heads. At 3600 rpm this is 8.33 msec.

3 Transfer. After the heads are positioned correctly according to **1** and **2**, the data is transferred at a rate proportional to the speed of the disc rotation and the density of recording on a given track. Typically this is of the order of 1–2 Mb per second.

The total access-time is the sum of these three effects. There are other delays, for example, the settle time for the heads to stop oscillating when they have stopped over the correct track, but these are insignificant compared with the above. In practical seismic data processing, the important time to minimise is **1**. This is normally achieved using a variety of techniques, the details of which are irrelevant here.

In the seismic industry, a typical computer system will have disc storage of between 600 and 40 000 Mb available.

1.2.4 Plotting and plotters

One of the most important aspects of a seismic computing system is that most users are only interested in the plot. Yet many machines, especially large ones, have no on-line plotting facilities, making the whole business of producing a plot frustrating and difficult. The use of magnetic tape is involved to transport files to be plotted to the off-line plotting device, often part of a separate empire.

There are basically two kinds of plotter.

1 Analogue plotter. Examples of this kind of plotter include pen-plotters of various kinds, which are electro-mechanical and continuously variable within a certain plotting area. Pen-plotters may be either flat-bed or drum. Such plotters are used only for the production of maps in the seismic industry.

2 Digital plotter. These are probably the most common kind of plotter in exploration seismology and are mainly used to produce seismic sections. Such a plotter is not continuous in operation and constructs a plot by drawing dots sufficiently densely that the overall picture is produced accurately. Common dot (the term pixel is often used, but this has a much wider meaning in image processing) densities used are 100, 200 and upwards per inch. Some digital plotters used in both the seismic industry and in LANDSAT satellite imagery can manage 2000 dots per inch using laser technology. To all intents and purposes however, a seismic plot looks continuous for dot densities exceeding about 300 dots per inch.

There are two main kinds of plotting, vector and raster. Both kinds of plotting can be done on either kind of plotter, although vector plotting is most suited to an analogue plotter and raster plotting is best done on a digital plotter.

Vector plotting implies that the plot consists of a series of vectors, i.e. commands to the plotter to draw a line from A to B, then B to C and so on. It is easy to imagine why this is particularly suited to an analogue pen-plotter where the pens physically move about drawing these vectors. Raster plotting is so-called because of its physical similarity to the way a television picture is constructed from raster scans. A raster is a line of discrete plot positions at each of which a dot can be drawn or not. Such dots are often drawn by a row of styli, electrostatically depositing ink along the raster line, one for each plot position. Fig. 1.6 indicates how a plot would come off a raster plotter. It is obvious that such a plotter could be used to simulate vector drawing by considering Fig. 1.7. Here, dots are turned on at the points where the vector intersects the rasters.

Note finally, that characters are drawn on such plotters by simulating the character set by a unique set of vectors for each character to be plotted. For example, Fig. 1.8 shows how the letter E might be drawn on both an analogue and a digital plotter.

1.2.5 Floating point accelerators and array processors

In seismic work, and scientific work generally, a very considerable amount of calculation is done in the computer. Such calculation is generally done on floating

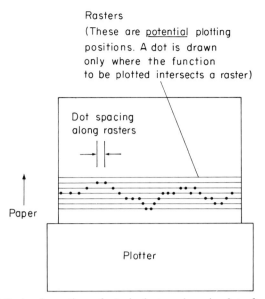

Fig. 1.6. A schematism of a 'wiggle-trace' mode plot of a seismic trace on a raster plotter.

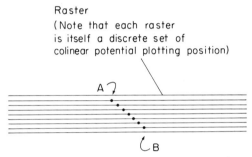

Fig. 1.7. A plot of a ' vector' (i.e. a line), joining points A and B on a raster plotter. Note that the higher the raster density (and correspondingly the dot density along a raster), the more the line of discrete dots resembles a continuous line.

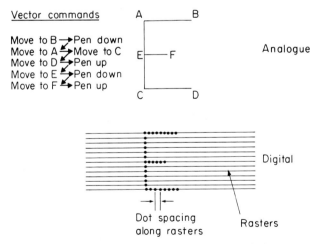

Fig. 1.8. Examples of the ways the letter 'E' might be produced on an analogue and a digital plotter.

point numbers, as described in the next section. It is a common requirement, therefore, to be able to do such floating point arithmetic as quickly as possible. This can be achieved by either:

1 Adding a special piece of hardware called a floating point accelerator to the CPU itself, if the computer manufacturer actually makes such a piece of equipment.

2 Adding a peripheral processor called an array processor, which will usually be made by a different manufacturer.

1 has the advantage in terms of the ease and speed of making the data available to the floating point arithmetic unit in that it usually shares main memory with the CPU and is invisible to the user's program. In the light of this, it might be wondered why **2** is an option at all. The answer seems to be an enlightened piece of business opportunity, taking advantage of the fact that the computer manufacturers were extremely slow (and still are) at providing high-speed floating point arithmetic at a reasonable price. Consequently, the array processor seems here to stay. Unfortunately, they do not make the job of geophysical computer program maintenance any easier and often confuse the algorithmic structure of a seismic data process, as will be seen shortly.

The unit of floating point performance is the megaflop which, believe it or not, stands for 1 million floating point instructions per second. It may also be used to describe certain Hollywood 'disaster' movies. A 10 megaflop array processor can out-perform almost all CPUs available on floating point arithmetic including those with floating point accelerators. Today's most powerful technology (based around something called a pipeline) is capable of sustaining anywhere between 100 and 800 megaflops at a price to match it. The competing technology is called parallelism and seems on physical grounds the only choice for the future, although still in its infancy.

Array processors are commonly driven by FORTRAN-callable subroutines which cause various kinds of floating point activity to take place. As was mentioned above however, they do confuse the overall algorithmic structure.

For example, consider the following FORTRAN program fragment which computes a well-known formula in exploration seismology:

```
      DO 100 I = 1, NSAMPS
            T(I) = SQRT( TZERO(I) ** 2 + (X/V(I)) ** 2 )
100 CONTINUE
```

The equivalent in a typical array-processor environment would read something like:

```
      CALL  PUTSCL(X,..)
      CALL  PUTVEC(V,..)
      CALL  PUTVEC(TZERO,..)
      CALL  DIVVEC(X,..,V,..)
      CALL  SQRVEC(TZERO,..)
      CALL  SQRVEC(X/V,..)
      CALL  ADDVEC(TZERO,..,X/V,..,SUM,..)
      CALL  ROOTVC(SUM,..)
      CALL  GETVEC(T,..)
      CALL  APCOMP
```

The details of each vector function are not relevant to this discussion. Simply note that the reduction in clarity and consequent room for error is very marked. Suffice it to say however, that without the development of fast floating point arithmetic, seismic data processing would still be in its relative infancy.

1.2.6 Internal data formats

Before finishing this chapter, it is worthwhile devoting a few words to the internal representation of information in the computer.

There are essentially three kinds of data format used commonly in the general seismic computing environment:

1 Integer. This consists of the value of an integer expressed in a binary form of whatever length (word length) is appropriate to the computer concerned. The first bit is almost universally a sign bit, although negative numbers may be expressed in either 1's complement or 2's complement form. The details are extremely tedious and do not generally affect the exploration geophysicist, but for anybody wishing to read about such sybaritic delights, most books on computer programming have pages devoted to the subject, effecting a harmless cure for insomnia.

2 Floating point. In this format, the computer word is split up into an exponent and a mantissa or fraction, in a scientific notation form. Multiplication of such numbers involves addition of the exponents and multiplication of the mantissae. This format has by far the greatest dynamic range and is consequently most suitable for scientific computation. The floating point format used on SEG Y tapes is that of IBM computers.

3 Character. On almost all machines, the ASCII (American Standard Code for Information Interchange) 7-bit code is used to express CHARACTER data, with one character per byte. Unfortunately, due to the predominance of one manufacturer, IBM, in the early days of seismic computing, the IBM 8-bit character format EBCDIC (Extended Binary Coded Decimal Interchange Code) is used as the character format of the SEG Y tape header.

Finally, to see how such numbers are used in the computer programs which are used to process seismic data, consider the following simple FORTRAN program:

```
INTEGER     INT
REAL        REL
CHARACTER   CHAR

INT = 1
REL = 1.
CHAR = '1'
```

If we dumped the contents of INT, REL and CHAR in a hexadecimal (or base 16) dump on a machine which used IBM data formats, it would be seen that

```
INT    would contain 00000001
REL    would contain 41100000
CHAR   would contain F1     (EBCDIC, remember)
```

Note that a hexadecimal dump uses the numbers 0–9 and the letters A–F to represent numbers 10–15. Hence, to interpret F1 above for example, F is the hexadecimal value denoting 15, corresponding to a bit pattern 1111 and 1 is hexadecimal for the bit pattern 0001. Hence F1 corresponds to eight bits or one byte as was described in Section 1.2.1. Similarly 41100000 corresponds to a 32 bit or four byte word.

Such dumps are often done when trying to identify odd field tape formats, as still occur in spite of the considerable efforts which have been made by the industry at standardisation.

Chapter 2
Time Series Analysis in Seismology

2.1 Introduction

Like many other sciences, that of time series analysis has experienced an enormous period of growth in the last twenty years following the advent of the digital computer. Originally an abstruse branch of statistics, it has grown to the point where it rivals other more traditional disciplines in scope, level of activity and especially in the amount of literature. Consequently, all that will be attempted here is to give the geophysicist a 'feel' for those parts of the subject of particular relevance to seismic data acquisition and processing, and to indicate some of the many references which give a more detailed exposition than is possible here.

For a comprehensive review and theoretical treatment of the development of time series analysis in seismology, Robinson and Treitel (1980) is highly recommended. For background reading and insight into the usage and analysis of time series in general, Chatfield (1975) gives a very palatable account of a subject which in some hands has proven rather indigestible.

What then is a time series? Quite simply, it is no more than a series of observations of some entity taken repeatedly throughout some period of time. Time series may be continuous or discrete, and if discrete, regularly or irregularly sampled. An example of an irregularly sampled discrete series is the current exchange rate. Although sampled daily during the week, it is not sampled at weekends in spite of the fact that events may occur which would affect it. As this book is depressingly full of seismic time series, the first example of a discrete, regularly sampled time series, as shown in Fig. 2.1, is a digital representation of a descriptive set of snow indicators published by Jackson (1977).

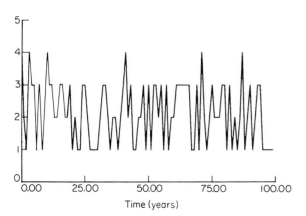

Fig. 2.1. UK snowfall indices for the 100 years prior to 1976. The time series is discrete-valued with four possible values of 1, 2, 3 and 4, with 4 representing the most snowy.

In both earthquake and exploration seismology generally, only discrete, regularly sampled time series are considered as recorded by modern digital seismic data acquisition equipment. After introducing the underlying concepts, this book will also restrict itself to discrete, regularly spaced time series.

What information can be gleaned from time series analysis? The answer is often a remarkable amount considering the degree of noise pollution which can occur. In seismology, time series are:

1 Analysed for significant periodicities.

2 Used to interpolate an observation at a point in time at which it was not recorded, using observations recorded both before and after the desired time.

3 Used to extrapolate or predict. Techniques are available which allow predictions to be made outside the period of time for which observations are available. This may be before they start, after they end or both.

4 Filtered deterministically to enhance or attenuate particular frequency components of known good or bad quality. Quantitative assessments of quality are available by measurements of something known as signal to noise ratio, of which much more later.

5 Filtered statistically to enhance or attenuate particular frequency components of unknown signal to noise ratio.

6 Inverted in some sense to yield information about the subsurface from observations made at the surface. This may take many forms, all of which involve fitting a model of some predetermined kind to the surface-recorded observations.

Exploration seismologists traditionally work in the band of frequencies between 0 and about 1000 Hz and usually in the low end of this band. Note that Hz is short for hertz or cycles per second and also that 0 Hz is commonly known as D.C. amongst seismologists, which itself is short for direct current, illustrating the electrical engineering ancestry of the pioneers of the subject. Two features single out seismological time series from those occurring in most other sciences. First, they are generally of very poor quality, especially considering the huge amount of money spent on the sole purpose of improving them. Second, this poor quality necessitates prodigious quantity to exploit statistical data processing techniques based on redundancy. A typical large seismic survey carried out in the marine environment may yield a dataset of around a million million bytes (see para. 1.2.1). Enormous operational problems are involved in handling this quantity of data. Also the sheer volume of the data usually precludes many of the sophisticated techniques which are available to other sciences and which could benefit raw seismic data greatly. The earth certainly does not give up its secrets easily, or at any rate, cheaply. Nevertheless, the final processed quality of seismic data has improved dramatically since its first appearance in digital form in the early 1960s.

In the form in which most seismic data processing algorithms are applied, a single time series appears as a series of numbers usually totalling between 1000 and 16 000 samples which represent particle displacement, particle velocity or pressure, measured every few milliseconds for several seconds, depending on the depth of the target. Exploration seismologists are rarely interested in any echoes which take longer than about 8 seconds to return, corresponding to their principal goal of hydrocarbon reserves. However, in some parts of the world, present-day technology is capable of picking up reflections comfortably beyond the Moho. The sampling interval in exploration seismology is usually 1, 2, or 4 milliseconds although 1/4 and 1/2 millisecond sampling intervals are used in high-resolution seismics to broaden the valid range of frequencies as, for example, when attempting to delineate coal measures. Whether or not there is any useful information at these frequencies is of course another matter and depends on many factors.

Fig. 2.2 shows an example of the first 4 seconds of a seismic time series recorded off the coast of Ireland at a sampling interval of 4 msec (or equivalently a sampling rate of 250 Hz). Beware, it is quite common to hear the expression '... a sampling rate of 4 msec ...', which as any pedant will tell you, is incorrect terminology. Note that the time series in Fig. 2.2 is shown in its raw form as it appears before any processing. Fig. 2.3 shows the first second of this seismic trace or record in more detail, whilst Fig. 2.4 shows the first 7 of the 96 traces

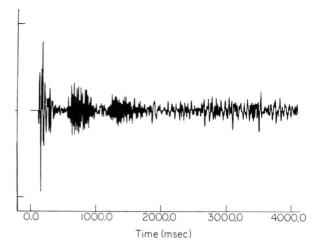

Fig. 2.2. The first 4 seconds of a near trace extracted from a typical marine shot file. The trace has been approximately compensated for decay using an exponential function.

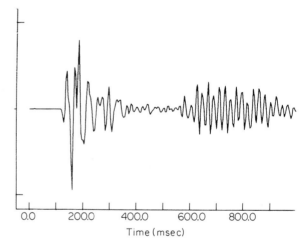

Fig. 2.3. The first second of the time series of Fig. 2.2.

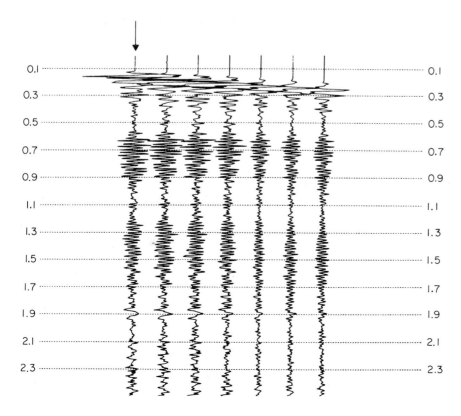

Fig. 2.4. The first 7 traces from the shot file from which the trace of Fig. 2.3 was extracted.

recorded for this particular shot. The time series shown in Figs. 2.2 and 2.3 is marked with an arrow. Note the increasing delay in the energy arrival from left to right. This corresponds to the increasing physical separation between the seismic source location (shot location) and the location of the corresponding receiver. The seismic survey geometry, or relative positions of the seismic source and receivers will be considered in much greater detail in Chapter 3.

As mentioned earlier, seismic time series are often studied for evidence of periodicities, and the natural domain in which to carry out this search is the frequency domain. Facility in this domain normally extends far beyond a knowledge of the more rigorous aspects of the various transformations involved, and the importance of a good working knowledge of the relationships between the time domain and the frequency domain cannot be underestimated. The next and subsequent sections will give a brief idea of Fourier theory and some of the insights it can provide as a supplement to the time domain viewpoint.

2.2 Basic Fourier theory

It is said that when Fourier originally stated his thesis that arbitrary functions could be represented by simple trigonometric series, the idea was categorically refuted by such famous mathematicians as Euler, Lagrange and d'Alembert (cf. Kline, 1972). Refreshingly, they were wrong. However, even Fourier's most ardent admirers could not have foreseen the extent to which his work would be applied as a result of the invention of the digital computer.

The Fourier transform occurs in three common forms for each of three classes of function:
1 Continuous periodic functions. Such functions repeat with some period T. In this case, the function is expressed as a discrete, infinite summation and the Fourier coefficients are expressed as integrals.
2 Continuous aperiodic functions. Such functions do not repeat. In this case both function and Fourier coefficients are expressed as integrals.
3 Discrete periodic and aperiodic functions. In this case, both function and Fourier coefficients are expressed as discrete summations.

The basic result of Fourier is expressed by case 1. Case 2 can be derived from case 1. As seismologists almost invariably have to deal with discrete time series, it is case 3 which will be of most interest here. In what follows therefore, cases 1 and 2 will be developed briefly for completeness, and then attention will be fixed on case 3 almost without exception in the rest of this chapter.

The basic result, case 1, is as follows. Suppose $x(t)$ is a periodic function of time t, with period T as shown in Fig. 2.5. Then x can be expressed as an infinite trigonometric series

$$x(t) = a_0 + 2 \sum_{k=1}^{\infty}$$
$$\cdot \left(a_k \cdot \cos\left(\frac{2\pi kt}{T}\right) + b_k \cdot \sin\left(\frac{2\pi kt}{T}\right) \right) \qquad (2.2.1)$$

where

$$a_k = \frac{1}{T} \int_0^T x(t) \cdot \cos\left(\frac{2\pi kt}{T}\right) dt \qquad k \geqq 0$$

$$b_k = \frac{1}{T} \int_0^T x(t) \cdot \sin\left(\frac{2\pi kt}{T}\right) dt \qquad k \geqq 1 \qquad (2.2.2)$$

Equations (2.2.1) and (2.2.2) are valid almost universally in the real world and will always be assumed to be so here. For more details on when they are not and indeed most subjects concerning the Fourier transform, the reader is referred to Bracewell (1978) for an extremely comprehensive account. For the rigorously minded, the original and definitive work which introduces the concept of the generalised function is Lighthill (1962).

The extension of the above to case 2, that of the continuous aperiodic function, can be accomplished heuristically as follows:

First, note that the kth coefficient a_k or b_k, corresponds to a conventionally defined circular frequency of

$$w_k = \frac{2\pi k}{T} \qquad (2.2.3)$$

and that the spacing between adjacent frequency components in the summation is

$$w_k - w_{k-1} = \frac{2\pi}{T} \qquad (2.2.4)$$

For aperiodic functions, T is infinite and letting T tend to infinity in (2.2.4), it can be seen that adjacent frequencies or harmonics as they are known merge together in the summation (2.2.1). Carrying the analysis through (see, for example, Bracewell (loc. cit.) or Newland (1975)), the equivalent result to that of (2.2.1–2.2.2) for aperiodic functions is

$$x(t) = \int_{-\infty}^{\infty} X(k) \cdot \exp\left(i2\pi kt\right) dk \qquad (2.2.5)$$

where

$$X(k) = \int_{-\infty}^{\infty} x(t) \cdot \exp\left(-i2\pi kt\right) dt \qquad (2.2.6)$$

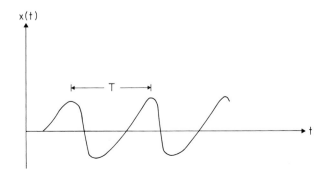

x(t)

Fig. 2.5. A continuous time function of period T.

As was mentioned above, this derivation has been included for completeness as the pragmatic approach taken here requires only results for discrete, evenly-spaced time series. To that end, the discrete form of the Fourier transform corresponding to case **3** will now be derived and developed.

2.3 The discrete Fourier transform

2.3.1 Derivation from the continuous case

To aid in deriving the discrete form of the equations (2.2.1) and (2.2.2) define

$$X_k \equiv a_k - ib_k \qquad (2.3.1)$$

(2.2.2) can then be written

$$X_k = \frac{1}{T} \int_0^T x(t) \cdot \exp\left(-i\left(\frac{2\pi kt}{T}\right)\right) dt \qquad (2.3.2)$$

Now consider what happens if the continuous time series $x(t)$ is known only at a discrete set of values of t, which are equidistant in t. Let $x(t)$ be the set $x_0, x_1, x_2, \ldots, x_{N-1}$. Here time t has been defined as $t = rd$, where d is the so-called sample interval and r is an integer in the range $(0, \ldots, N-1)$. Obviously,

$$d = T/N \qquad (2.3.3)$$

The integral (2.3.2) can then be written

$$X_k = \frac{1}{T} \sum_{r=0}^{N-1} x_r \cdot \exp\left(-i\left(\frac{2\pi krd}{T}\right)\right) \cdot d \qquad (2.3.4)$$

This is equivalent to assuming that the area under the continuous curve shown in Fig. 2.6 is replaced by the sum of the shaded strips.

Finally, putting $T = Nd$ in (2.3.4) gives

$$X_k = \frac{1}{N} \sum_{r=0}^{N-1} x_r \cdot \exp\left(-i\left(\frac{2\pi kr}{N}\right)\right) \qquad (2.3.5)$$

This may then be considered the discrete equivalent to be used in calculating the coefficients in the Fourier series (2.2.1) as given by (2.2.2).

Two matters of great importance should be noted about the discrete Fourier transform (DFT) given in (2.3.5):

1 The derivation leading to (2.3.5) involved a discrete approximation to an integral. The DFT is therefore an approximation which may be expected to improve as N gets very large and the sample interval d, therefore, gets

correspondingly small. For small N, the approximation to the continuous series at points other than the discrete evenly-spaced values $x, \ldots$ may not be very good, an oft forgotten fact.

2 The DFT (2.3.5) does allow the discrete evenly-spaced values $x_0, \ldots$ to be regained exactly. Any value x_s of the series $x_0, x_1, \ldots$ is given exactly by the inverse formula

$$x_s = \sum_{k=0}^{N-1} X_k \cdot \exp\left(i\left(\frac{2\pi ks}{N}\right)\right) \qquad (2.3.6)$$

To see this, using (2.3.5),

$$\sum_{k=0}^{N-1} X_k \cdot \exp\left(i\left(\frac{2\pi ks}{N}\right)\right)$$

$$= \frac{1}{N} \sum_{k=0}^{N-1} \sum_{r=0}^{N-1} x_r \cdot \exp\left(-i\left(\frac{2\pi k(r-s)}{N}\right)\right)$$

Change the order of summation to get

$$= \frac{1}{N} \sum_{r=0}^{N-1} x_r \cdot \left[\sum_{k=0}^{N-1} \exp\left(-i\left(\frac{2\pi k(r-s)}{N}\right)\right)\right]$$

Now k, r, s, and N are all integers, so

$$\left[\qquad\right] = \begin{array}{l} N \text{ if } s = r \\ 0 \text{ otherwise} \end{array}$$

and so

$$\frac{1}{N} \sum_{r=0}^{N-1} x_r \cdot \left[\sum_{k=0}^{N-1} \exp\left(-i\left(\frac{2\pi k(r-s)}{N}\right)\right)\right] = x_s$$

For reasons which will be apparent shortly, the following definition will be used.

The formal definition of the DFT of a series x_r, where $r = 0, 1, 2, \ldots, N-1$ is

$$X_k = \frac{1}{N} \sum_{r=0}^{N-1} x_r \cdot \exp\left(-i\left(\frac{2\pi kr}{N}\right)\right) \qquad (2.3.7)$$

$$k = 0, 1, \ldots, N-1$$

and the inverse DFT (IDFT) is given by

$$x_r = \sum_{k=0}^{N-1} X_k \cdot \exp\left(i\left(\frac{2\pi kr}{N}\right)\right) \qquad (2.3.8)$$

$$r = 0, 1, \ldots, N-1$$

Complete symmetry may be achieved by using $N^{-1/2}$ in both (2.3.7) and (2.3.8).

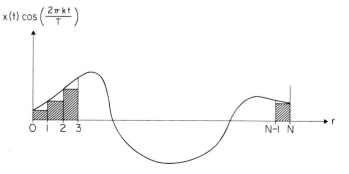

Fig. 2.6. This diagram illustrates the inherent approximation of the DFT compared to the continuous Fourier transform.

2.3.2 DFT relationships

Periodicity

The DFT is by definition periodic with period N. Setting $k = N + l$ in (2.3.7) leads directly to

$$X_{N+l} = X_l \qquad (2.3.9)$$

Real time domain series

Many time series met in exploration seismology are real. Hence taking the complex conjugate of (2.3.5) gives

$$
\begin{aligned}
\bar{X}_k &= \frac{1}{N} \sum_{r=0}^{N-1} x_r \cdot \exp\left(i\left(\frac{2\pi kr}{N}\right) \right) \\
&= \frac{1}{N} \sum_{r=0}^{N-1} x_r \cdot \exp\left(i\left(\frac{2\pi(-k)r}{N}\right) \right) \\
& \qquad k = 0, 1, \ldots, N-1 \\
&= X_{-k}
\end{aligned}
$$

from which one can deduce

$$\bar{X}_k = X_{-k} \qquad (2.3.10)$$

It can immediately be deduced from (2.3.9) and (2.3.10) that

$$\bar{X}_0 = X_0 \qquad (2.3.11)$$

Hence the zero frequency value (also commonly known as D.C.) is real and since

$$X_N = X_0 \qquad (2.3.12)$$

sample N is also real by virtue of the periodicity. Finally, by using (2.3.9) and (2.3.10), the following result can be proved

$$\bar{X}_{N/2} = X_{N/2} \qquad (2.3.13)$$

Hence sample $N/2$ is also real. In general, (2.3.9) and (2.3.10) give $\bar{X}_k = X_{N-k}$. Fig. 2.7 summarises these symmetries. Sample $N/2$ corresponds to the so-called Nyquist frequency and plays a central role in the discretisation of continuous time series as will now be seen.

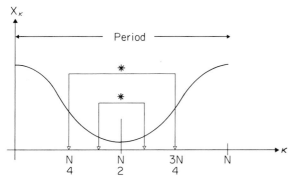

Fig. 2.7. The conjugate symmetry of the Fourier transform of a real time series. Note that * here denotes complex conjugate.

2.3.3 Sampling and aliassing

As has been intimated, there is a further subtlety associated with the act of sampling. Using (2.3.9) it may be seen that only the values X_k, $k = 0, 1, \ldots, N-1$ are unique and furthermore, using (2.3.10)

$$|X_k| = |X_{-k}|$$

Hence only that part of $|X_k|$ for which k is less than or equal to $N/2$ is unique. Using (2.2.3), this corresponds to a circular frequency range of

$$|w| \leqq \frac{2\pi N/2}{T} = \frac{\pi}{d} \qquad (2.3.14)$$

where d is the time domain sampling interval. Any frequencies originally in the signal before discretisation which are greater than the limit set by (2.3.14) are distorted and manifest themselves as frequencies within the above limit by virtue of being folded or wrapped-round the Nyquist circular frequency π/d.

The corresponding Nyquist linear frequency sampling limitation is given by

$$|f| \leqq \frac{1}{2\pi} \cdot \frac{\pi}{d} = \frac{1}{2d} \qquad (2.3.15)$$

This is a most important result and will be referred to throughout this work as the Whittaker–Nyquist–Shannon (WNS) criterion, acknowledging the pioneering contributions of these three authors.

The distortion induced when this limitation is not satisfied is known as 'aliassing'. For example, in exploration seismology, the commonly used sample interval of 4 msec causes frequencies greater than

$$\frac{1}{2(.004)} = 125 \text{ Hz}$$

to be aliassed. An original frequency of 150 Hz present in the original signal before discretisation will appear in the discretised version folded around the Nyquist at

$$125 - (150 - 125) = 100 \text{ Hz}$$

The only solution to this problem is to perform analogue filtering on the original signal before it is discretised, in order to limit the original bandwidth appropriately. The subject of aliassing will be raised again shortly when (2.3.15), which is true only for signals of infinite duration, is extended for the physically realistic case of finite duration. This extension also admits of a particularly simple way of visualising how aliassing occurs in terms of convolution, as will be discussed later.

One final point to note is that the Nyquist limitation is a bandwidth limitation. It is perfectly possible to sample a signal which contains frequencies between 1000 and 1125 Hz unambiguously using a 4 msec sample interval, provided that it is known that no frequencies outside this band are present. This allows economy of sampling under certain special conditions.

Finally, since the degree of aliassing depends on the rate at which the input spectrum tends to zero, it is well worthwhile quoting the following useful result. If N is the highest order continuous derivative in the time domain, then the corresponding spectrum tends to zero as f^{-N}. Hence a highly continuous time domain function like an exponential curve has a spectrum which tends to zero very rapidly.

2.3.4 Time shifts and the DFT

Consider now the effect on X_k of a time shift of m units on the time domain series x_r. Rewriting (2.3.7)

$$X_k = \frac{1}{N} \sum_{r=0}^{N-1} x_r \cdot \exp\left(-i\left(\frac{2\pi kr}{N}\right)\right)$$ (2.3.16)

$$k = 0, 1, \ldots, N-1$$

The time-shifted version can be written

$$X_k' = \frac{1}{N} \sum_{r=0}^{N-1} x_{r+m} \cdot \exp\left(-i\left(\frac{2\pi kr}{N}\right)\right)$$ (2.3.17)

$$k = 0, 1, \ldots, N-1$$

How are X and X' related? First, set $r' = r + m$. Then (2.3.16) and (2.3.17) together imply

$$X_k' = \frac{1}{N} \sum_{r'=1}^{N-1+m} x_{r'} \cdot \exp\left(-i\left(\frac{2\pi kr'}{N}\right)\right)$$

$$\cdot \exp\left(i\left(\frac{2\pi km}{N}\right)\right)$$

Splitting the summation up into two parts, from m to $N-1$ and from N to $N-1+m$ and using the equality

$$x_{N+r'} = x_{r'}$$

the following result is obtained

$$X_k' = \exp\left(+i\left(\frac{2\pi km}{N}\right)\right) \cdot X_k$$ (2.3.18)

Hence, if X_k is the transform of x_r, then

$$\exp\left(+i\left(\frac{2\pi km}{N}\right)\right) \cdot X_k$$

is the transform of x_{r+m}.

Consider now the effect on the amplitude and phase spectra of X_k. By definition,

$$X_k = A_k \cdot \exp\left(i2\pi P_k\right)$$ (2.3.19)

where A_k is the amplitude and P_k is the phase spectrum, then

$$X_k' = A_k \cdot \exp\left(+i2\pi\left(\frac{km}{N} + P_k\right)\right)$$ (2.3.20)

Hence, the time shift of m units (possibly fractional) may be performed in the frequency domain by adding the linear ramp km/N to the phase spectrum of X_k. The amplitude spectrum is unaffected. It can easily be checked that (2.3.10) to (2.3.12) are satisfied by this provided that the Nyquist frequency value $X_{N/2}$, is zero. A non-zero Nyquist frequency value is indicative of alias-ing, although a zero value does not imply that aliassing is absent.

2.3.5 The fast Fourier transform (FFT)

No description of the DFT would be complete without some mention of the fast Fourier transform or FFT as it is universally abbreviated. A large number of books already contain as much or as little as the reader could wish to know on the subject. Here a very brief description indeed will be attempted with some emphasis on practical details. For an authoritative account, Oppenheim and Schafer (1975) or Newland (1975) are both highly recommended.

As its name would suggest, the FFT is simply an elegant way of performing DFTs quickly. The existence of this algorithm makes the DFT a practical tool as, without it, even very powerful computers would be insufficiently fast to perform DFTs on large time series.

The basic DFT requires a number of operations of the order of N^2 to complete. The FFT, on the other hand, requires only $N \log_2 N$.

Even for a typical FFT length in exploration seismology of 4096 samples, the ratio of DFT to FFT time is greater than 340. In addition, the greatly reduced number of operations ensures greater accuracy because of a correspondingly slower growth of round-off errors resulting from the finite-length arithmetic available in a computer. A computer with a floating point arithmetic capability of the order of 10 megaflops, or 10 million floating point instructions per second is capable of doing such a FFT in a few milliseconds.

How then does the algorithm work? To see this consider a time series

$$x_r \qquad r = 0, 1, \ldots, N-1$$

Now define the following two half-sequences

$$y_r = x_{2r}$$
$$\qquad r = 0, 1, \ldots, (N/2 - 1)$$
$$z_r = x_{2r+1}$$

It will be shown that the DFTs of the two half-sequences are intimately related to the DFT of the full sequence. It should be noted that this ordering of elements in the two sequences is known as decimation in time. (An alternative ordering known as decimation in frequency whereby the two sequences are composed of adjacent elements may also be used. The name arises from the fact that adjacent ordering in the time domain leads to an alternate or decimated ordering in the frequency domain.) The DFTs of y and z are

$$Y_k = \frac{1}{N/2} \sum_{r=0}^{N/2-1} y_r \cdot \exp\left(-i\left(\frac{2\pi kr}{N/2}\right)\right)$$

$$k = 0, 1, \ldots, N/2 - 1$$

$$Z_k = \frac{1}{N/2} \sum_{r=0}^{N/2-1} z_r \cdot \exp\left(-i\left(\frac{2\pi kr}{N/2}\right)\right)$$

$$k = 0, 1, \ldots, N/2 - 1$$

Suitable manipulation reveals that

$$X_k = \frac{1}{2}\left(Y_k + \exp\left(-i\left(\frac{2\pi k}{N}\right)\right) \cdot Z_k \right)$$

$$k = 0, 1, \ldots, N/2 - 1$$

$$X_{k+N/2} = \frac{1}{2}\left(Y_k - \exp\left(-i\left(\frac{2\pi k}{N}\right)\right) \cdot Z_k \right)$$

These last two equations form the computational 'butterflies', so-called because of a passing resemblance to lepidoptera when displayed in graphical form!

This splicing together of two half-sequences is known as a radix-2 algorithm and it demands that N is a power of 2. Other radix algorithms exist but their additional complication precludes further discussion here. The radix-2 algorithm proceeds iteratively taking N one-point transforms, which it then splices using the above butterflies to produce $N/2$ 2-point transforms. These are then spliced in turn to produce $N/4$ 4-point transforms and so on until the complete transform is left.

A number of other clever speed-ups are used such as bit-reversing to index consecutive elements of the half-sequences in place, but the detail is unnecessary here and the reader is referred to the books by Oppenheimer and Schafer or Newland cited earlier.

Finally, the prospective user should note the following:

1 The number of points in the time series must be a power of 2, which is normally achieved in practice by padding with zeroes. In some circumstances, such as computing convolutions (see later), such padding is essential to produce a correct result.

2 To save space when transforming a real time series, a number of proprietary algorithms make use of the symmetry in the frequency domain to pack the transform into the same space as the real series. For example, if there are N real time samples, N complex samples would normally result, taking twice as much space in a computer. However, as (2.3.10) to (2.3.13) showed, there are only $N/2 + 1$ unique values in the transform. Since it is known that the imaginary parts of the D.C. component and the Nyquist are zero, the real part of the Nyquist is placed into the imaginary part of the D.C. component producing a slightly scrambled transform, but which fits exactly into the same space as the original time series. Potential users should make sure that their transform is of the unscrambled variety before further use!

2.3.6 An example of FFT usage: interpolation

The example of interpolation has been used because it is a little off the beaten track in terms of FFT usage and has some points of interest.

Interpolation is very extensively used in the processing of seismic data, and the performance of interpolators is of considerable importance. The importance of the FFT lies in the fact that it admits of a perfect interpolator, the yardstick against which all other kinds of interpolators may be measured. Interpolation is of

sufficient importance to be considered in detail later in this chapter after the concept of spectra has been introduced. The reason for this deferral is that the quality of an interpolator can be judged very simply in the frequency domain, a domain which seismologists so often have to resort to, when deciding data processing parameters. For now, it will be noted simply that the point of interpolation is to determine the values of a discretely sampled time series at points other than those at which it is specified, providing those points lie between the minimum and maximum specified time.

Having built up the relevance of the FFT in interpolation, there is of course a caveat. The FFT may only be used to interpolate samples half-way between existing samples, that is the sample interval may only be reduced by a power of two. This subdivision could be repeated indefinitely to achieve any required interpolation, but the economics are unattractive in practice.

As an example, consider Table 2.3.1. The left-hand column contains an eight sample discretised sinusoid of period equal to eight times the sample interval. The samples are written in (real, imaginary) pairs. The right-hand side contains the FFT of this sinusoid in the format indicated in Fig. 2.7. The aim is to super-sample this time series by a factor of two, thereby interpolating values of the sinusoid half-way between the points at which it was specified. Sixteen samples will result.

Table 2.3.1

Input sinusoid	FFT
(1. , 0.)	(0., 0.)
(0.71, 0.)	(8., 0.)
(0. , 0.)	(0., 0.)
(−0.71, 0.)	(0., 0.)
(−1. , 0.)	(0., 0.)
(−0.71, 0.)	(0., 0.)
(0. , 0.)	(0., 0.)
(0.71, 0.)	(8., 0.)

In order to interpolate this, 8 complex zeroes are inserted at the old Nyquist sample position (row 4 in the right-hand column), thus maintaining the appropriate symmetry. The result, and the inverse transform containing the interpolated sinusoid, are shown in Table 2.3.2 below. Note that the spectrum is essentially unaltered.

It is very important to note that if the input time series is aliassed, i.e. it possesses a non-zero Nyquist value, nonsense proportional to the degree of aliassing will be generated. Time domain methods of interpolation are generally much less sensitive to the effects of aliassing.

2.3.7 An example of DFT usage: the spectrum of a ghost

Although spectra will be considered in more detail in Section 2.9, it is worthwhile at this stage to consider a time-domain response of great importance to exploration seismologists as an example of an analytic DFT.

Table 2.3.2

Interpolated sinusoid		FFT	
	(1. , 0.)	(0., 0.)	
	− (0.92, 0.)	(8., 0.)	
	(0.71, 0.)	(0., 0.)	
	− (0.38, 0.)	(0., 0.)	
	(0. , 0.)	(0., 0.) −	
	− (− 0.38, 0.)	(0., 0.) −	
	(− 0.71, 0.)	(0., 0.) −	
Generated	− (− 0.92, 0.)	(0., 0.) −	Inserted
samples	(− 1. , 0.)	(0., 0.) −	complex zeroes
	− (− 0.92, 0.)	(0., 0.) −	
	(− 0.71, 0.)	(0., 0.) −	
	− (− 0.38, 0.)	(0., 0.) −	
	(0. , 0.)	(0., 0.)	
	− (0.38, 0.)	(0., 0.)	
	(0.71, 0.)	(0., 0.)	
	− (0.92, 0.)	(8., 0.)	

The time domain response to be considered is known as the ghost and it appears in all recorded seismograms simply by virtue of the nature of seismic data acquisition. It is considered in more detail in Section 2.4 and also in Chapter 3, but for now it will be sufficient to explore some of its properties, starting from the following definition. The response may be defined as

$$
\begin{aligned}
x_m &= 1 & m &= n \\
&= -1 & m &= n + p \\
&= 0 & &\text{otherwise}
\end{aligned}
\tag{2.3.21}
$$

Using (2.3.5), the DFT of this function is, after some manipulation

$$
X_k = \frac{2}{N} \exp\left(- i\left(\frac{2\pi k[n + p/2]}{N} - 1\right)\right)
$$
$$
\cdot \left[\sin\left(\frac{\pi k p}{N}\right)\right]
\tag{2.3.22}
$$

Significantly, this can attain zero whenever

$$
\frac{kp}{N} = j \qquad j = 0, 1, 2 \ldots
\tag{2.3.23}
$$

For $j = 1$

$$
k = \frac{N}{p}
\tag{2.3.24}
$$

Now, from the earlier discussion $k = N$ in the DFT corresponds to $f = 1/d$ in hertz, where d is the sample interval and $1/d$ is simply twice the Nyquist frequency. Hence, $k = N/p$ corresponds to $f = 1/dp$, leading to the very important result

$$
f_j = \frac{j}{\text{Time delay}} \qquad j = 0, 1, 2 \ldots
\tag{2.3.25}
$$

where f_j is the frequency at which the spectrum of the ghost is zero.

The reader might like to repeat the above calculation

for a ghost function which has $-a$ instead of -1 for $(0 < a < 1)$.

Figs. 2.8 and 2.9 show typical ghost spectra for a delay of 10 msec with $a = 1$ and 0.8 respectively.

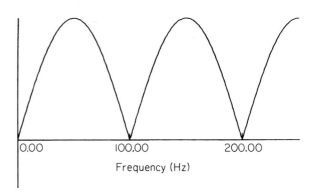

Fig. 2.8. The amplitude spectrum of a 'ghost' time series consisting of $(1, 0, 0, 0, 0, -1)$ with a sampling interval of 2 msec. An often forgotten fact is that the D.C. value is zero as well as the harmonics of main ghost notch (at 100 Hz in this case).

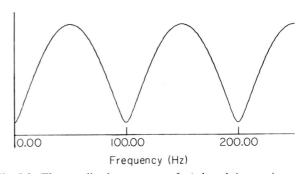

Fig. 2.9. The amplitude spectrum of a 'ghost' time series consisting of $(1, 0, 0, 0, 0, -.8)$ with a sampling interval of 2 msec.

2.4 Convolution

2.4.1 The 1-D convolutional model of the seismic trace

Convolution is a mathematical operation which combines two functions together in a certain way to produce a third. It is of great importance in electrical engineering as it is the mechanism by which linear systems combine their responses. It is of equally great importance in seismology as it appears that under a fairly wide range of circumstances, the earth behaves convolutionally.

The complete transient convolution, s_k, of two discrete functions given by

$$
\begin{aligned}
w_i \qquad i &= 0, 1, \ldots, L_w \\
e_j \qquad j &= 0, 1, \ldots, L_e
\end{aligned}
$$

is defined to be

$$
s_k = \sum_{j=0}^{L_w} w_j e_{k-j}
\tag{2.4.1}
$$

where $k = 0, 1, \ldots, L_s = L_w + L_e - 1$.

It is conventionally written as

$$s = w * e \qquad (2.4.2)$$

where $*$ denotes convolution.

The convolution may be thought of as a super-position of a scaled, shifted copy of w for each sample of e.

In seismology, the w-function is normally identified as the impinging seismic source wavelet and e a time series containing the earth's acoustic impedance function. Acoustic impedance is simply the product of density and the compressional velocity of sound, and it is the variations in this function which lead to echoes reflecting back up to the surface. The function e is consequently known as the reflection series. The resulting seismogram is denoted as s. Fig. 2.10 illustrates this concept schematically. In any real seismic experiment, additive noise is inevitably present leading to the so-called 1-D convolutional model of the seismic trace.

$$s = w * e + n \qquad (2.4.3)$$

where n denotes the additive noise.

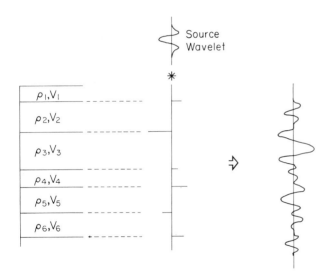

Fig. 2.10. The relationship between the earth's acoustic impedance function and an ideal seismic trace.

Before going on to discuss this further, a theorem of great practical and theoretical importance will now be proved.

2.4.2 The discrete convolution theorem

The reason that this result is so important is that it provides a way of computing long convolutions accurately and efficiently. It is of great theoretical importance because of the insight that it gives into the mechanism of convolutional filtering.

Again, let S be the DFT of s
W be the DFT of w
E be the DFT of e.

Suppose also that we are considering the noise-free case, so

$$s = w * e$$

or, discretely

$$s_k = \sum_{j=0}^{L_w} w_j e_{k-j} \qquad k = 0, 1, \ldots, L_s$$

where $\quad L_s = L_w + L_e - 1 \qquad (2.4.4)$

Using (2.3.8),

$$s_k = \sum_{l=0}^{N-1} S_l \cdot \exp\left(i\left(\frac{2\pi kl}{N}\right)\right) \qquad (2.4.5)$$

$$w_j = \sum_{m=0}^{N-1} W_m \cdot \exp\left(i\left(\frac{2\pi jm}{N}\right)\right) \qquad (2.4.6)$$

$$e_{k-j} = \sum_{n=0}^{N-1} E_n \cdot \exp\left(i\left(\frac{2\pi(k-j)n}{N}\right)\right) \qquad (2.4.7)$$

Using (2.4.5)–(2.4.7) and exchanging orders, (2.4.4) can be written

$$\sum_{l=0}^{N-1} S_l \cdot \exp\left(i\left(\frac{2\pi kl}{N}\right)\right)$$
$$= \sum_{m=0}^{N-1} W_m \sum_{n=0}^{N-1} E_n \cdot I \qquad (2.4.8)$$

where

$$I = \sum_{j=0}^{L_w} \exp\left(i\left(\frac{2\pi jm}{N}\right)\right) \cdot \exp\left(i\left(\frac{2\pi(k-j)n}{N}\right)\right)$$
$$= N \cdot \exp\left(i\left(\frac{2\pi kn}{N}\right)\right) \cdot D_{n-m} \qquad (2.4.9)$$

and

$$D_{n-m} = 1 \qquad n = m$$
$$= 0 \qquad \text{otherwise}$$

using the orthogonality of the exp functions.

This reduces (2.4.8) to

$$\sum_{l=0}^{N-1} S_l \cdot \exp\left(i\left(\frac{2\pi kl}{N}\right)\right)$$
$$= N \cdot \sum_{m=0}^{N-1} W_m \cdot E_m \cdot \exp\left(i\left(\frac{2\pi km}{N}\right)\right) \qquad (2.4.10)$$

Multiplying both sides of (2.4.10) by $\exp(-i(2\pi kp/N))$ and summing from $p = 0$ to $N - 1$ and again using the orthogonality of the exp functions gives the final result

$$S_p = N \cdot W_p \cdot E_p \qquad p = 0, 1, \ldots, N - 1 \quad (2.4.11)$$

(2.4.11) is the discrete expression of the convolution theorem, which states that convolution in the time domain is equivalent to multiplication of the respective Fourier transforms in the frequency domain.

Note that the factor N appearing in (2.4.11) is a consequence of the definition of the DFT, (2.3.7) and (2.3.8). Had the $1/N$ been included on the inverse transform rather than the forward transform, it would not have been present. Alternatively, had $N^{-1/2}$ been used in the DFT itself, the factor would also have been $N^{-1/2}$.

2.4.3 The convolution theorem and circularity

In practice, when convolutions are calculated using the convolution theorem (i.e. the DFTs of the two series to be convolved are taken, multiplied, and the result transformed back to the time domain), it is most important to make sure that N, the length of the DFT, equals or exceeds the length of the complete transient convolution. This may be expressed by

$$N \geqq L_e + L_w - 1 \qquad (2.4.12)$$

If this does not hold, a phenomenon known variously as circularity or wrap-round will occur, due entirely to the fact that usage of the DFT in the form given in (2.3.7) and (2.3.8) implies that the time domain series are periodic with period N samples. The problem is illustrated in Fig. 2.11.

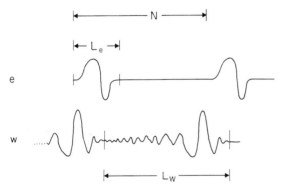

Fig. 2.11. Convolutional wrap-round due to circularity.

The reader may reasonably ask 'Why bother doing it in the frequency domain?'. The answer purely and simply is a mixture of economics and precision.

As far as economics is concerned, this issue has already been discussed earlier in connection with the FFT. The result is every bit as important for convolution. Fig. 2.12 shows timing comparisons between convolution in the time domain and its equivalent in the frequency domain as exemplified by (2.4.11).

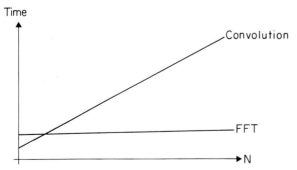

Fig. 2.12. The relative execution time of convolution compared to frequency domain multiply as a function of N, the length of the time series, for a given filter length.

To quantify this graph, note that a convolution carried out using (2.4.1) takes a number of operations proportional to

$$L_e \cdot L_w \qquad (2.4.13)$$

Using the FFT and (2.4.11), the total number of operations is approximately

$$3N \cdot \log_2 N + N \qquad (2.4.14)$$

corresponding to three transforms and one vector multiply, where N is the next power of two, exceeding or equalling the sum of L_e and L_w. In normal usage in exploration seismology L_e is approximately 3000, in which case convolution is quicker in the frequency domain if L_w is greater than about 100, although this is very hardware dependent.

From the point of view of precision, it is important to note that it is empirically observed that round off errors can accumulate rapidly for long filters of the order of two hundred samples or more when these filters are applied convolutionally in the time domain. This is of course dependent on the word length of the machine in question, but the fact remains that the situation is much better in the frequency domain simply because the number of operations is so much smaller.

2.4.4 Extended 1-D convolutional model of the seismic trace

In Section 2.4.1, a very restrictive form of the convolutional model was considered. In practice, many other effects appear. Consider then, the following extended model:

$$s(t) = w(t) * e(t) * s_g(t) * r_g(t) * i(t)$$
$$+ n(t) \qquad (2.4.15)$$

where w is the seismic source wavelet
e is the earth's reflection series
s_g is the source 'ghost'
r_g is the receiver 'ghost'
i is the instrument response.

These effects will be discussed in much more detail in Chapter 3, but the point of introducing them here is to show that the convolution theorem admits of a very simple interpretation in the frequency domain, viz.

$$S(f) = W(f) \cdot E(f) \cdot S_g(f) \cdot R_g(f) \cdot I(f)$$
$$+ N(f) \qquad (2.4.16)$$

where the transforms of the time domain functions are simply denoted by upper case. The responses are multiplicative which allows the geophysicist to have a much better intuitive feel for the concatenation of the various filtering effects.

Note the following with respect to equation (2.4.16):
1 If the source or receiver ghost are zero at any frequency, then the whole response S is either pure noise or zero also, depending on the presence or absence of noise.

2 S is what we actually measure. Ideally the geophysicist would like E. In practice, W is generally unknown and the noise is always unknown. This is a classic example of the inverse problem as discussed in much greater detail in Chapter 5.

3 The form of (2.4.16) suggests that it is possible to remove known convolutional responses from S simply by frequency domain division, provided that the known responses are not singular in which case a little judicious fiddling may be necessary. This indeed is the case and the technique is known as deconvolution. This is a vitally important subject to seismologists and will be discussed at great length later when an important method of achieving it in the time domain is introduced.

2.4.5 Review of assumptions

It may be helpful at this stage to review some of the results and the assumptions made in the various continuous and discrete definitions so far:

1 For continuous time series, convolution in the time or frequency domain is equivalent to multiplication in the other domain. No assumptions are made about periodicity.

2 For discrete time series, convolution in the time or frequency domain is equivalent to multiplication in the other domain provided the time series are periodic. This assumption is required for the proof to work.

3 Linear convolution as occurs in the earth is defined as in equation (2.4.1) without any assumptions made about the participating functions.

4 Circular convolution is linear convolution where the participating input functions are periodic.

5 Circular convolution in the time or frequency domain is the same as multiplication in the other for discrete series.

6 Circular convolution agrees with linear convolution at those lags for which the two input series do not overlap. Hence linear convolution can be achieved cheaply by padding with zeroes, forward transform, multiplication in the 'other domain', inverse transform and extraction of those lags which do not incur circularity.

In the next section, filtering will be introduced.

2.5 Filtering

2.5.1 Introduction

Filtering may be defined as the act of modifying a time series by application of another time series of some kind, in some characteristic way.

In seismology, filtering is nearly always taken as meaning multiplicative modification of the spectrum of the seismogram (and so by the convolution theorem discussed in the last section, modification by convolution in the time domain). Note that on its own, filtering says nothing about the kind of filtering which has been or is to be applied. The word must be further qualified,

for example, band-pass filtering, filtering by the recording instruments, anti-alias filtering and so on.

A distinction must also be made between data-dependent and data-independent filtering. In data-dependent filtering, the characteristics of the filter are derived from the data to be filtered, for example, predictive deconvolution discussed in Section 2.8. In data-independent filtering, as will be considered here, the filter characteristics are not derived directly from the data (although a geophysicist will generally choose the specification of such a filter to be compatible with some property of the data to be filtered).

Finally, note that in this section, attention will be restricted to that class of filters which are designed and applied at the data processing stage as digital filters, rather than the analogue filtering which takes place both in the earth itself, and in the recording equipment.

As was mentioned above, the application of the filter is equivalent to a convolution of the filter with the seismogram. It is intuitively much simpler however, to consider the action of the filters in the frequency domain:

Let S' be the DFT of the filtered seismogram
F be the DFT of the filter
S be the DFT of the unfiltered seismogram

then

$$S'_k = N \cdot F_k \cdot S_k \tag{2.5.1}$$

as was shown in the previous section. In this section, the properties of the frequency domain filter coefficients, F_k will be under scrutiny. Before proceeding, the following terms should be defined.

Decibels

A non-dimensional unit of scale. Suppose 2 numbers, X and Y, are of different magnitudes but in the same units and both greater than 0. Define

$$D = 20 \cdot \log_{10}\left(\frac{Y}{X}\right) \tag{2.5.2}$$

Then if Y is bigger than X, X is D db (decibels) down on or in comparison to X. The decibel originates in acoustics and its logarithmic definition reflects the fact that the human ear reacts logarithmically to sound. Example:

If $Y = 2X$, then D is approximately 6 db.

Octave

An octave is a doubling of frequency. For example, 120 Hz is an octave above 60 Hz and so on.

2.5.2 The band-pass filter

The spectrum of a typical band-pass filter is shown in Fig. 2.13.

The filter has unit amplitude response between frequencies corresponding to p and q. This is known as the

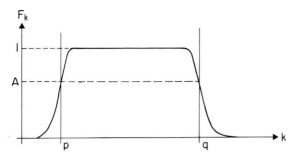

Fig. 2.13. The amplitude spectrum of a typical band-pass filter. A denotes the half-power point and p and q the cut-off frequencies.

pass-band. The frequencies are normally defined by the points at which the amplitude of the filter reaches A, where A is 3 db down on the level within the pass-band. (In actuality, this description results from the electrical engineering ancestry of such filters. Electrical engineers typically consider the half-power point as the edge of the pass-band. As will be seen, half-power corresponds to the square-root of two of the amplitude which is very close to 3 db. These kind of filters arise naturally in the study of RCL (resistance, capacitance, inductance networks).) On either side of the pass-band, the response rapidly approaches zero but, for a finite time-duration filter, can never reach it. The slopes of the filter response are discussed later in this section. Obviously, multiplication by such a filter will restrict the pass-band of the seismogram in (2.5.1) to p and q also. In exploration seismology, the choice of p and q is made by a processing geophysicist, usually subjectively, and is based on the distribution of signal and noise in the data. For example, a filter with a pass-band of 10–70 Hz might be used in the shallower part of the section, corresponding to two-way times of less than 1 or 2 seconds.

2.5.3 High-pass filters

The spectrum of a typical high-pass (also known as low-cut) filter is shown in Fig. 2.14.

Here, p is the low-cut or cut-off frequency. A is as defined for a band-pass filter. Application of such a filter according to (2.5.1) would restrict all frequencies in the seismogram S, to be greater than that corresponding to p. Note again that frequencies below that corresponding to p do not disappear, they are just

multiplicatively reduced. That they cannot disappear is a function of the fact that only an infinitely long time domain signal can have a true cut-off frequency, that is a frequency beyond which the spectrum is zero.

2.5.4 Low-pass filters

The spectrum of a typical low-pass filter (also known as a high-cut) is shown in Fig. 2.15.

Here, q is the high-cut or cut-off frequency. Again A is as defined for the band-pass filter. Application of this filter according to (2.5.1) will reduce all frequencies above that corresponding to q by the appropriate amount.

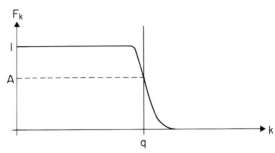

Fig. 2.15. The amplitude spectrum of a typical low-pass filter. A denotes the half-power point and q the cut-off frequency.

Note the following:
1 A band-pass filter can be constructed as the product of a suitable low-pass and high-pass filter.
2 The reason band-pass filters were discussed first is that they are by far the most commonly used filters in exploration seismology. This is because of the fact that, in seismic data, signal to noise ratios tend to decrease both for very low frequencies and high frequencies. This subject is discussed at length in Section 2.7.

2.5.5 Notch filters

This is a rather specialist filter and its spectrum typically looks like that shown in Fig. 2.16.

Here m is the notch frequency. Its action is obvious. It is often used in land seismic data processing for the suppression of power-line interference, for example, 50 Hz and its harmonics in the UK and 60 Hz and its harmonics in the USA. A harmonic is simply an integer multiple of the basic frequency.

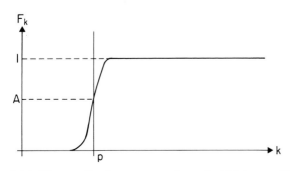

Fig. 2.14. The amplitude spectrum of a typical high-pass filter. A denotes the half-power point and p the cut-off frequency.

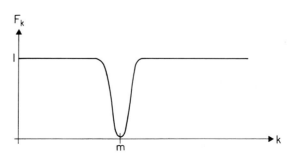

Fig. 2.16. The amplitude spectrum of a typical notch filter, where m denotes the notch frequency.

2.5.6 The slopes of a filter

The slopes of a filter are important in that it is essential to know how much a certain frequency will be reduced by application of the filter. For example, in anti-alias filtering, where frequencies of greater than a certain amount (the Nyquist frequency appropriate to the new sampling interval) must be removed before resampling to a coarser sample interval, the filter slopes are important insofar as they cannot be infinite and so must start to act before the Nyquist frequency in order to guarantee sufficient attenuation at the Nyquist itself. So, how should they be specified? There are two methods in common usage:

1 Specify the actual amplitude values at four different frequencies, the inner two of which correspond to the pass-band.
2 Specify the frequencies corresponding to the pass-band, then specify the slopes as being in db per octave. For example, if a low-pass filter has a slope of 24 db per octave starting at a cut-off frequency of 60 Hz, the amplitude spectrum of the filter will be 24 db down on its 60 Hz value at 120 Hz. (Note that the definition of the pass-band implies that it will be $24 + 3 = 27$ db down on the pass-band value.) A typical band-pass filter might then be 10–70 Hz pass-band with a low slope of 12 db per octave and a high slope of 72 db per octave. It should be noted that real filters always finish up with a somewhat non-linear slope in the db/octave domain, so the slope should be taken as a nominal value only.

2.5.7 Ringing

A practical phenomenon to be guarded against when designing filter slopes is that of ringing. Ideally, the amplitude spectrum of a band-pass filter ought to look something like Fig. 2.17. If, however, the inverse DFT is

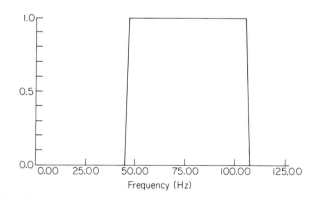

Fig. 2.17. The amplitude spectrum of an ideal band-pass filter.

taken, assuming a zero phase spectrum, the impulse or time domain response shown in Fig. 2.18 results. An obvious property of this impulse response is its oscillatory nature away from the central peak. This is the phenomenon known as ringing which occurs whenever filter slopes are too steep. Closer inspection reveals that the ringing is a superposition of the frequencies at each end of the pass-band as the following analysis shows.

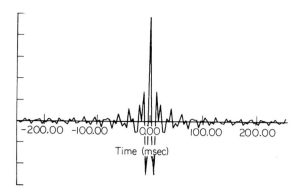

Fig. 2.18. The equivalent time domain response of the filter of Fig. 2.17. Note the oscillations (ringing) away from the central peak caused by the steepness of the slopes in the amplitude spectrum.

Assume the amplitude spectrum is given by

$$S_k = 1 \qquad L \leq k \leq M$$
$$\quad = 0 \qquad \text{otherwise}$$

then the inverse transform is

$$S_r = \sum_{k=L}^{M} \exp\left(i\left(\frac{2\pi k r}{N}\right)\right) \qquad r = 0, 1, \ldots, N-1$$

this is a simple geometric series whose sum is

$$S_r = \frac{\exp\left(i\left(\frac{2\pi L r}{N}\right)\right) - \exp\left(i\left(\frac{2\pi (M+1) r}{N}\right)\right)}{1 - \exp\left(i\left(\frac{2\pi r}{N}\right)\right)}$$

$$(2.5.3)$$

This is the desired result. S_r is the difference of two sinusoids of frequencies corresponding to the edges of the pass-band, modulated by the factor in the denominator. Note that $M + 1$ rather than M arises because of the discrete nature of the transform.

To see that the difference of the two sinusoids dies away slowly, observe that the denominator can be written as

$$-i\frac{2\pi r}{N}$$

for $r \ll N$, corresponding to a $1/t$ decay.

In order to reduce ringing to acceptable levels, the slopes must be reduced in some way. In practice, a quarter cycle of a $\cos^2$ function of an appropriate frequency is normally used, but even a simple linear taper as shown in Fig. 2.19 has a dramatic effect, as shown in the corresponding impulse response in Fig. 2.20. This can be understood from another viewpoint in light of the knowledge that the linear taper in Fig. 2.19 was achieved by convolution with another rectangular or 'box-car' function in the frequency domain. As has already been seen, this corresponds to multiplication in the time domain by a function which decays also as $1/t$ away from the peak value, hence reducing the ringing.

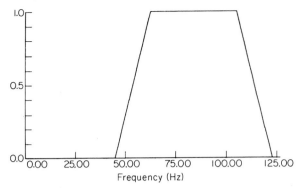

Fig. 2.19. The amplitude spectrum of a typical band-pass filter.

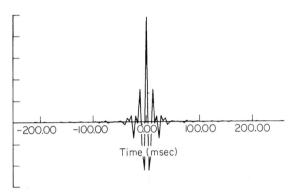

Fig. 2.20. The equivalent time domain response of the filter of Fig. 2.19. Note the lack of oscillations (ringing) away from the central peak, as a result of the shallower slopes.

Having discussed practical filtering in some detail, further insight into the theoretical nature of filtering will now be sought.

2.6 The Z-transform, phase and delay

2.6.1 The Z-transform

Consider the following time series with evenly spaced samples in time

$$\ldots 1, 2, 0, -1, 0, 1, \ldots \qquad (2.6.1)$$

and construct the following polynomial in z

$$P(z) = \ldots 1 + 2z - z^3 + z^5 \ldots \qquad (2.6.2)$$

This is called a Z-transform.

Note that if (2.6.2) is multiplied by z, the effect is exactly the same as if the original time series had been shifted along by one time sample. Because of this phenomenon, z is called the 'unit-delay operator'.

From this it may be seen that multiplication by z^n delays the time series by n time units, whereas multiplication by z^{-n} advances the time series by n units.

The Z-transform is intimately related to the Fourier transform, as will now be shown. Define

$A(z)$ as the Z-transform of the time series a_t
$B(z)$ as the Z-transform of the time series b_t
$C(z)$ as the Z-transform of the time series c_t

so

$$A(z) = \sum a_t \cdot z^t$$
$$B(z) = \sum b_t \cdot z^t \qquad (2.6.3)$$
$$C(z) = \sum c_t \cdot z^t$$

Now consider the product of two polynomials as follows:

$$C(z) = A(z)B(z) \qquad (2.6.4)$$

Substitution of (2.6.3) in (2.6.4) and equating powers of z gives

$$c_0 = a_0 b_0$$
$$c_1 = a_1 b_0 + a_0 b_1$$
$$c_2 = a_2 b_0 + a_1 b_1 + a_0 b_2$$

and in general,

$$c_k = \sum_{i=0}^{L} a_i \cdot b_{k-i} \qquad k = 0, 1, \ldots \qquad (2.6.5)$$

where L is the length of a.

This is, however, the convolution equation of Section 2.4 (equation (2.4.1)). Hence, multiplication of the Z-transforms is equivalent to convolution of the original time series. But, recalling the convolution theorem (equation (2.4.11)), so is multiplication of the Fourier transforms. Hence, the Fourier transform of a time series may be formally identified as a special case of the Z-transform when

$$z = \exp\left(-i\left(\frac{2\pi i l}{N}\right)\right) \qquad (2.6.6)$$

for any $l = 0, 1, \ldots, N - 1$

So, $C(z)$ can be interpreted as either a time domain function, by identifying coefficients of powers of z, or as a frequency domain function by using (2.6.6).

It is fair to wonder why the Z-transform has been introduced at all at this stage. The reason is that, amongst other things, the Z-transform admits of a very convenient way of discussing phase.

2.6.2 Minimum, mixed and maximum phase and dipoles

Consider the simple time series of length 2

$$a + bz \qquad (2.6.7)$$

This can be written as (a, b) which is known as a dipole.

Departing temporarily from the discrete domain, consider the continuous Fourier transform of this time series, which can be achieved by interpreting z as

$$z = \exp(-iw) \qquad (2.6.8)$$

where w is the angular frequency.

Then (2.6.7) can be written as

$$|A(w)| \cdot \exp(iP(w))$$

where

$$|A(w)| = [a^2 + 2ab \cdot \cos(w) + b^2]^{1/2} \qquad (2.6.9)$$

is the amplitude spectrum and

$$P(w) = \tan^{-1}\left(\frac{-b \cdot \sin(w)}{a + b \cdot \cos(w)}\right) \qquad (2.6.10)$$

is the phase spectrum.

The following should be noted:

1 It is obvious from (2.6.9) that (a, b) and (b, a) have the same amplitude spectrum.

2 If $a > b$, then the phase of (a, b) is everywhere less than the phase spectrum of (b, a). For this reason, if $a > b$, (a, b) is known as the 'minimum-phase dipole' and (b, a) is known as the 'maximum-phase dipole'.

3 The earliest time sample a, of the minimum-phase dipole is bigger than that of the maximum-phase dipole, b.

Now consider the Z-transform of the wavelet $(6, 5, 1)$. This

$$= 6 + 5z + z^2$$
$$= (3 + z) \cdot (2 + z)$$
$$= (3, 1) \cdot (2, 1)$$

A fundamental theorem of algebra tells us that any polynomial of degree n can be factored into n polynomials of order 1 (the zeroes of the polynomial). Such a wavelet is defined to be minimum phase if and only if each of its constituent dipoles is minimum phase (since phase is additive for the dipoles, you get the minimum only if they are all minimum).

Note also that if any combination of dipoles have their constituent elements swapped, e.g. (a, b) to (b, a), the composite amplitude spectrum is unchanged, but the phase is changed. This leads to the idea that for a wavelet of length N samples, there are 2^N wavelets all of the same amplitude spectrum, but with different phase spectra varying between the minimum and maximum. Only one of these has the minimum-phase property. There is also only one maximum-phase wavelet, corresponding to the time reverse of the minimum-phase wavelet, which is equivalent to flipping all the constituent dipoles round to put their smallest component first.

Why then is the concept of minimum phase of interest? There are two reasons.

1 Of all wavelets with the same amplitude spectrum which are zero before time zero, the one with minimum phase has the shortest time duration, which makes it important from the point of resolution. This property of being zero before time zero is also known as the property of causality.

2 Certain other processing procedures perform at their best when the wavelet is minimum phase, notably, predictive deconvolution, as is discussed in Section 2.8.

2.6.3 Partial energy and minimum delay

The pth partial energy of a wavelet b_t is defined as

$$E_p = \sum_{t=0}^{p} b_t^2 \qquad (2.6.11)$$

It can be shown, cf. Robinson and Treitel (1980), that for the minimum-phase wavelet, this function is larger at any p than for any other wavelet with the same amplitude spectrum. When discussing energy in this way, the concept is known as 'minimum delay'. Put another way, for the minimum-phase wavelet, the energy arrives soonest. It is instructive to note as a corollary, that the first sample of the minimum-phase wavelet must be bigger than the first sample of any other wavelet of the same amplitude spectrum, simply because when decomposing the wavelet into its constituent dipoles, the coefficient of z^0, i.e. the first sample, is the product of all the first components of the dipoles. These are of course the bigger in each dipole since each of the dipoles themselves must be minimum phase in order that the composite wavelet be minimum phase.

2.6.4 The equivalent minimum-phase wavelet and the Hilbert transform

As has been seen, the minimum-phase wavelet of a family of wavelets of the same amplitude spectrum is of great interest to the processing geophysicist. The equivalent minimum-phase wavelet to a given wavelet is simply that wavelet which has the minimum-phase spectrum but the same amplitude spectrum as the given wavelet.

The geophysicist is often faced with the problem of constructing the equivalent minimum-phase wavelet for a given wavelet. There are several ways of doing it:

1 From the Z-transform, decompose the wavelet into dipoles. Flip the maximum-phase ones round. Multiply the dipoles together. Extracting coefficients of the powers of z in the resulting Z-transform gives the required minimum-phase wavelet. Numerically, this requires finding the N complex roots of a polynomial of order N. In practice, for long wavelets, this does not prove very satisfactory, but the method is by far the easiest to understand.

2 The inverse-inverse method. A method depending on the Wiener inverse filter. The Wiener filter is discussed in Section 2.8, but for its usage in the context here, the reader is referred to Chapter 3 of Claerbout (1976), for a detailed discussion.

3 The Hilbert transform method. In practice, probably the most often used, although its rather profound relationship with the concepts of minimum phase and causality precludes further discussion until a later chapter. The reader is also referred to Kanasewich (1981) for detailed background or again to Claerbout (1976) for operational details, including a computer program. The transform itself is, however, very simple and will be discussed shortly in connection with the analytic signal.

2.6.5 Zero-phase wavelets and minimum-time duration

It was stated earlier that the minimum-phase wavelet had the shortest time duration of all causal wavelets with the same amplitude spectrum. It does not, however, have the shortest duration of all wavelets with this amplitude spectrum. This is a property of that wavelet of the suite which has a zero-phase spectrum. This wavelet is, by virtue of its phase spectrum, symmetric about time-zero. It is thus non-causal and in the ideal case, extends to infinity on both sides of time-zero. There is however nothing wrong with non-causal wavelets in data after processing, when all the data is to hand and 'time-zero' is flexible. Indeed, a conversion to zero-phase is very often done using appropriate shaping-filters, (cf. Section 2.8), so that this property of minimum-time duration may aid the processing and analysis of the data.

Whilst on this subject, this seems an appropriate place to correct a widely believed fallacy. It is quite true that the convolution of a minimum-phase wavelet with another minimum-phase wavelet results in a new minimum-phase wavelet with an amplitude spectrum which is the product of the amplitude spectra of the input wavelets. It is not true, however, that the convolution of a minimum-phase wavelet with a zero-phase wavelet also results in a minimum-phase wavelet. The fallacy arises in assuming that because phase is additive then the original minimum-phase spectrum is unaltered and therefore remains minimum phase. This is only true if the amplitude spectrum of the original minimum-phase wavelet is also unaltered, which is tantamount to saying that the zero-phase wavelet does nothing as it has a wider bandwidth than the minimim-phase wavelet. In practice, however, the amplitude spectrum of the minimum-phase wavelet is altered and the resulting wavelet is mixed-phase.

To see this, suppose that $A_M(f)$, $P_M(f)$, $A_Z(f)$, $P_Z(f)$ and $A_R(f)$, $P_R(f)$ are the amplitude and phase spectra of a minimum-phase wavelet, its equivalent zero-phase wavelet, and the result of convolving these two respectively. $P_Z(f) = 0$ by definition. Then

$$A_M(f) \cdot A_Z(f) = A_R(f)$$
$$P_M(f) = P_R(f)$$

But the equivalent zero-phase wavelet, by definition, has the same amplitude spectrum as the original minimum-phase wavelet. So

$$A_M(f) \cdot A_M(f) = A_R(f)$$
$$P_M(f) + P_M(f) = P'_R(f)$$

Now, P'_R must be the minimum-phase spectrum corresponding to A_R, because it results from the convolution of two minimum-phase wavelets (actually the same one twice).
    But

$$P_M(f) \neq 0, \qquad (P_Z(f) = 0)$$

Hence

$$P'_R(f) \neq P_R(f)$$

So, $P_R(f)$ cannot be the minimum-phase spectrum for this wavelet as the minimum-phase spectrum is unique for a given wavelet. Furthermore, it certainly isn't zero, hence it must be mixed-phase.

The above argument is illustrated in Figs. 2.21 to 2.24. Fig. 2.21 shows a minimum-phase wavelet with a 15–45 Hz pass-band with low and high slopes of 12 and 36 db respectively. The wavelet consists of 32 samples of 4 msec data. Fig. 2.22 shows the equivalent zero-phase wavelet. Figs. 2.23 and 2.24 are the convolutions of the minimum-phase wavelet with itself and with its zero-phase equivalent respectively, both shifted by 52 msec to show the energy which arrives before 'time-zero' in the latter case. They are dramatically different.

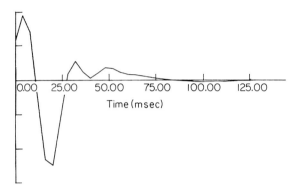

Fig. 2.21. The time domain response of a minimum-phase band-pass filter with a pass-band of 12–45 Hz and a low slope of 15 db per octave and a high slope of 36 db per octave.

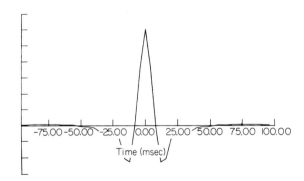

Fig. 2.22. The time domain response of the zero-phase filter with the same amplitude spectrum as that of Fig. 2.21.

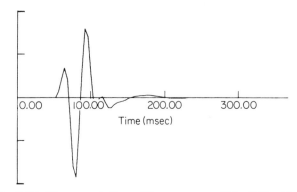

Fig. 2.23. The convolution of the response of Fig. 2.21 with itself, with a time shift of 52 msec for comparison purposes.

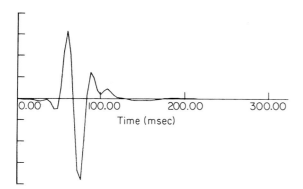

Fig. 2.24. The convolution of the responses of Figs. 2.21 and 2.22. The only difference between this response and the one shown in Fig. 2.23 is one of phase.

2.6.6 Causality and inversion

Consider the simple dipole (a, b) again. Now form its Z-transform.

$$a + bz \qquad (2.6.12)$$

A very common aim in exploration seismology is to invert some given response. This is done by designing some filter which when convolved with the given response produces a single spike at time-zero. The process is known as 'deconvolution'. Since multiplication of the respective Z-transforms is equivalent to convolution, the problem of inverting (2.6.12) for example, amounts to finding a Z-transform, $X(z)$, say, which when multiplied with (2.6.12) gives unity, resulting in deconvolution. So

$$X(z) \cdot (a + bz) = 1 \qquad (2.6.13)$$

remembering that unity is the Z-transform of a spike at zero-time. Obviously the unknown response is given by

$$X(z) = \frac{1}{a + bz} \qquad (2.6.14)$$

This can be expanded in powers of z as

$$X(z) = \frac{1}{a} \cdot \left[1 - \left(\frac{b}{a}\right) \cdot z + \left(\frac{b}{a}\right)^2 \cdot z^2 - \cdots \right]$$

$$(2.6.15)$$

Now, for $b < a$, i.e. a minimum-phase dipole, terms in this infinite series rapidly become very small. Hence this series could be truncated and still produce a filter which would invert (2.6.12) as accurately as is desired. In the time domain, this filter is

$$\left[\frac{1}{a}, -\frac{b}{a^2}, \frac{b^2}{a^3}, \cdots \right] \qquad (2.6.16)$$

If this filter is convolved with (2.6.12), a spike at zero-time would result. If a finite set of terms of this infinite series is convolved with (2.6.12), an approximation to a spike results. Such a filter is known as an all-memory component filter, because it only has non-zero coefficients for times after time-zero. Note that this truncated filter does not produce the 'best' inverting filter of a

given length. This topic is raised again in Section 2.8 where the Wiener filter will be discussed.

It is prudent to ask what happens if $b > a$, i.e. the dipole is maximum phase. Unfortunately, the infinite series (2.6.15) diverges, which implies that the inversion relies infinitely heavily on filter coefficients infinitely far off in the past, which is obviously wrong.

Mathematically, this problem can be overcome by expanding $X(z)$ as

$$X(z) = \frac{1}{bz\left(1 + \dfrac{a}{bz}\right)}$$

$$= \frac{1}{bz}\left[1 - \frac{a}{bz} + \left(\frac{a}{b}\right)^2 \cdot \left(\frac{1}{z}\right)^2 - \cdots \right] \qquad (2.6.17)$$

This filter behaves in an orderly way, with its coefficients rapidly decreasing in magnitude as they get further away from time-zero. However, they correspond to times before time-zero because of the negative powers of z. As such, this filter is said to be an all anticipation component filter. Such a filter is obviously non-causal but, as is mentioned above, can easily be constructed in the processing centre where all data past and future (within limits!) is at hand. In fact, filters possessing a mixture of memory and anticipation components (coefficients corresponding to times after and before time-zero respectively) are frequently used in processing to invert a wavelet which is mixed-phase. Such wavelets are not uncommon in exploration seismology.

2.6.7 The analytic signal

As the concept is a little difficult to grasp intuitively, the various relevant functions will first be defined. After this, their properties will be investigated and finally, their usage in the processing of seismic time series will be discussed in the light of these properties.

Let u_t be a time series with a corresponding Z-transform $U(z)$.

Consider now a filter h_t, with Z-transform $H(z)$, which, when applied to some input, shifts the phase of all frequency components by 90 degrees. The filter $H(z)$ is known as a Hilbert transformation filter.

Furthermore, let v_t, and its corresponding Z-transform $V(z)$, be the Hilbert transform of $U(z)$, i.e. the result of applying the Hilbert transform filter to u_t. Then

$$V(z) = H(z) \cdot U(z) \qquad (2.6.18)$$

Now define the analytic signal or complex trace of $U(z)$ as

$$A(z) = U(z) + i \cdot V(z) \qquad (2.6.19)$$

and define the energy envelope of $U(z)$ as

$$Y(z) = [\bar{U}(z) \cdot U(z) + \bar{V}(z) \cdot V(z)]^{1/2} \qquad (2.6.20)$$

where the bar as usual denotes the complex conjugate.

What are the properties of these various functions? The most important property from the point of view of the geophysicist is the treatment of phase in an input signal by these functions.

Consider an input

$$u_t = \cos(wt + P) \qquad (2.6.21)$$

where P is the phase and w is the circular frequency. Then, by definition,

$$v_t = \cos\left(wt + P + \frac{\pi}{2}\right)$$

$$= -\sin(wt + P) \qquad (2.6.22)$$

Substituting (2.6.21) and (2.6.22) in (2.6.20) gives

$$y_t = 1 \qquad (2.6.23)$$

where y_t is here defined to be the time domain function correspondent with $Y(z)$.

For a slightly more complicated input signal

$$u_t = \cos(w_1 t + P_1) + \cos(w_2 t + P_2) \qquad (2.6.24)$$

some manipulation results in

$$y_t = [2 + 2 \cdot \cos((w_1 - w_2)t + (P_1 - P_2))]^{1/2} \qquad (2.6.25)$$

This result has a most important interpretation. If an arbitrary phase is added to all components of the input signal, u_t, it does not appear in y_t because of the appearance of $P_1 - P_2$. Hence y_t is independent of phase shifts in u_t.

This is a very desirable property in seismology since a central objective of seismic interpretation is to measure the travel time of key seismic reflections, and continual phase changes in a seismogram can make this very difficult to do. Indeed, some geophysicists define the time of a reflecting horizon to be the time of the peak of the energy envelope, arguing that this is the most obvious thing to do.

A few years ago, it became popular to calculate various related functions, for example, writing the analytic signal (2.6.19) in the time domain form as

$$a_t = y_t \cdot \exp(ip_t) \qquad (2.6.26)$$

The instantaneous phase, p_t, may be written as

$$p_t = \frac{i}{y_t} \cdot \log(a_t) \qquad (2.6.27)$$

and the instantaneous frequency, f_t may be written as the derivative of (2.6.27), i.e.

$$f_t = \frac{dp_t}{dt} = i \frac{y_t}{a_t} \cdot \frac{d}{dt}\left(\frac{a_t}{y_t}\right) \qquad (2.6.28)$$

On a practical note, this latter function is normally calculated using a finite difference approximation to the differential coefficient.

Whilst these functions do provide alternative and sometimes valuable clues in the interpretation of seismic data, cf. Taner et al. (1979), it is probably fair to say that their usage has not been as widespread as it might have been due to their somewhat esoteric nature.

In the next section, two of the most vital functions in the processing of seismic time series will be considered in detail, both on their own merits and because of their intimate relationship with the Wiener filter.

2.7 The auto- and cross-correlation functions

2.7.1 Definitions and usage

The auto- and cross-correlation play a central part in the study of time series in general and of exploration seismology in particular. In general, cross-correlation functions are used to give a quantitative estimate of the degree of similarity between two time series as a function of a relative time shift between them. The auto-correlation function is simply a special case of this and measures the degree of similarity between a time series and a shifted copy of itself as a function of that shift.

Before attempting to define these functions, it should be noted that the following difference of nomenclature is sometimes encountered.

Seismology	Statistics
Correlation	Covariance
Normalized-correlation	Correlation

This occurs for both the auto- and cross-correlation functions. Since this is a book on seismology, the former definitions will be used.

Correlation functions are defined in terms of time series of infinite extent, but since the only concern here is for real experiments, the sample correlation functions will be used.

Suppose N samples of two time series x_t, y_t are available. Then the discrete auto-correlation r_k of a time series x_t is defined here by

$$r_k(x) = \frac{1}{N} \sum_{t=0}^{N-k-1} \bar{x}_t \cdot x_{t+k}$$

$$k = 0, 1, \ldots, N-1 \qquad (2.7.1)$$

where the $\bar{\ }$ denotes the complex conjugate and k is known as the lag.

Similarly, the discrete cross-correlation g_k of the time series x_t and y_t is defined here by

$$g_k(x, y) = \frac{1}{N} \sum_{t=0}^{N-k-1} \bar{x}_t \cdot y_{t+k}$$

$$k = 0, 1, \ldots, N-1 \qquad (2.7.2)$$

Note that in (2.7.1) and (2.7.2), the time series x_t and y_t have been assumed to have zero mean. In practice, the sample mean is computed and removed from the two series before the correlation functions are calculated. As usual, it will be assumed that the time series are real so that

$$\bar{x} = x$$

$$\bar{y} = y$$

Note the following important points:

1 For real, periodic time series of period N,

$$r_k = r_{-k} \tag{2.7.3}$$

2 Several subtle effects arise from assumptions about x_j and y_j when $j < 0$ or $> N - 1$, or allowing N to tend to infinity, when the summations (2.7.1) and (2.7.2) are allowed to extend from 0 to $N - 1$, as is sometimes the case. Restricting the summations to run up to $N - k$ is equivalent to assuming the time series are zero when $j < 0$ or $> N - 1$. Refer to the discussion in Section 2.4.5 for more details. For now it will simply be noted that the assumptions, in practice, depend on the usage to which the auto-correlation is put. When used for Wiener filter design, the time series is normally assumed to be zero outside the window of interest. For fairly large N, as is often the case, this assumption has only a small effect. When used for spectral analysis, the auto-correlation is often redefined to take into account this assumption for smaller values of N as

$$r_k(x) = \frac{1}{N-k} \sum_{t=0}^{N-1-k} \bar{x}_t \cdot x_{t+k}$$
$$k = 0, 1, \ldots, N-1 \tag{2.7.4}$$

although as a practical note, this is only valid for $k <$ about $N/10$. For k greater than this, the resulting auto-correlation may not be valid. Noting that correlation is simply convolution with one time series time-reversed, definitions (2.7.1) and (2.7.2) above are used because they allow the intimate relationship between convolution in the time domain and multiplication in the frequency domain, as discussed in Section 2.4, to be applied. This requires the assumption that the time series x and y are periodic with period N samples, and is one of the subtle effects referred to above.

3 The auto-correlation (2.7.1) is often normalised in practice such that its zero-lag value is 1, and is then known as the normalised auto-correlation. (It is a property of auto-correlations that the zero-lag value cannot be exceeded by any other lag value.)

4 If the time series is periodic with a certain period, its auto-correlation is periodic with the same period. This is proved in Section 2.7.3.

5 Such periodicities are easier to see in the auto-correlation than in the underlying time series if the time series is noisy. To see this, suppose that the time series can be written

$$x_t = s_t + n_t$$

where s_t is the signal and n_t is the additive noise at time index t.

Then the auto-correlation is

$$r_k(x) = \frac{1}{N} \sum_{t=0}^{N-1} (s_t + n_t) \cdot (s_{t+k} + n_{t+k})$$

Using some results from elementary statistics, and

assuming that the noise is zero mean and of standard deviation d,

$$E(s_t \cdot n_{t+k}) = E(s_{t+k} \cdot n_t) = 0$$
$$E(n_t \cdot n_{t+k}) = d^2 \qquad k = 0$$
$$= 0 \qquad \text{otherwise}$$

where $E(\ )$ is the statistical expectation operator, cf. Hogg and Craig (1959). Hence,

$$E(r_k(x)) = \frac{1}{N} \sum_{t=0}^{N-1} s_t \cdot s_t + d^2 \qquad (k = 0)$$
$$= \frac{1}{N} \sum_{t=0}^{N-1} s_t \cdot s_{t+k} \qquad (k > 0)$$

So, the only contribution from truly random noise is at the zero-lag of the auto-correlation, $k = 0$. Any periodicities manifesting themselves in the input signal should therefore be much more evident at non-zero-lags of the auto-correlation function. This is dramatically shown in Figs. 2.25–2.29. Fig. 2.25 is a simple 20 Hz sinusoid. Fig. 2.26 is its auto-correlation, assuming that the series is zero outside the limits shown. Note that the gradual decay in the auto-correlation function which this causes is the very trend which the factor $1/N - k$ is attempting to compensate for in equation (2.7.4). Fig. 2.27 is a Gaussian noise series of zero mean and unit variance

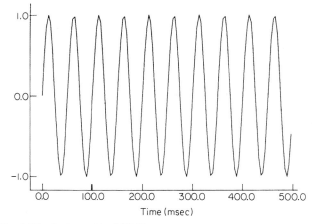

Fig. 2.25. A sinusoid of 20 Hz.

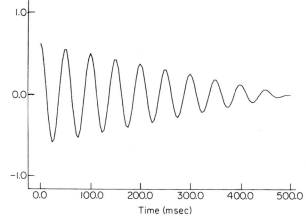

Fig. 2.26. The auto-correlation of the sinusoid shown in Fig. 2.25.

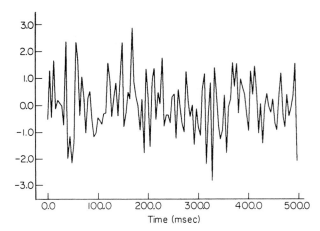

Fig. 2.27. A time series consisting of pure noise with a Gaussian distribution of zero mean and unit variance.

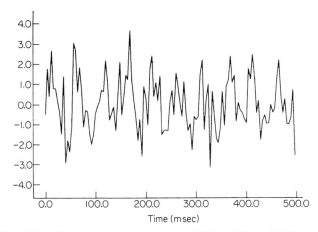

Fig. 2.28. The sum of the time series of Figs. 2.25 and 2.27. The S/N is approximately 0.5.

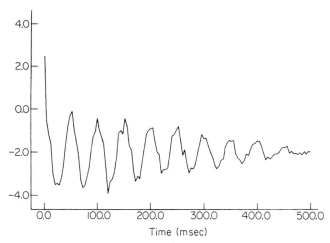

Fig. 2.29. The auto-correlation of the noisy sinusoid of Fig. 2.28. Note the clarity with which the inherent periodicity can now be seen and also the significant contribution of noise to the zero-lag position.

which, when added to the sinusoid of Fig. 2.25, results in the noisy sinusoid shown in Fig. 2.28. The sinusoidal energy in this series is virtually invisible, especially its period. Fig. 2.29 is the auto-correlation of Fig. 2.28. As is predicted by theory, the noise effectively manifests itself only at the zero-lag value. The underlying periodicity can clearly be seen and estimated.

This property of the auto-correlation is often exploited in practical seismic data processing, for example in choosing the prediction distance for a multiple attenuation deconvolution process. The inherent multiple periodicity is much easier to see in the auto-correlation function. This subject will be considered in depth in Chapter 3.

6 Both the auto-correlation and the cross-correlation functions appear as a central part of the theory of predictive deconvolution and Wiener filter theory as will be seen in Section 2.8.

The cross-correlation function affords an excellent and very widely used method of comparing two time series for likeness in the presence of noise. This is shown in Figs. 2.30–2.32. Figs. 2.30 and 2.31 show the same wavelet with considerable and different added noise, shifted in time with respect to each other. Fig. 2.32 shows the cross-correlation of these two time series. The obvious peak in the cross-correlation function shows that at a shift equal to the time of this peak, the two time series correlate well, as indeed they should. The point being that the time-shift is easier to choose just as the underlying periodicity in Fig. 2.29 was easier to choose. This is considered in more detail in Section 2.7.4.

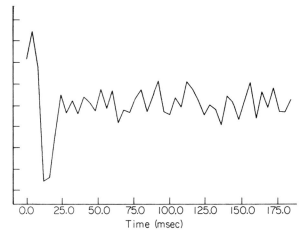

Fig. 2.30. An unshifted minimum-phase wavelet with additive noise. The general properties are unimportant.

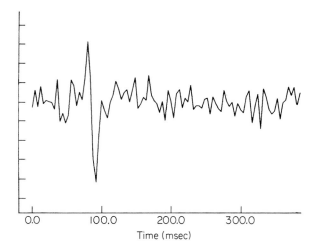

Fig. 2.31. The time series of Fig. 2.30 time-shifted by about 80 msec.

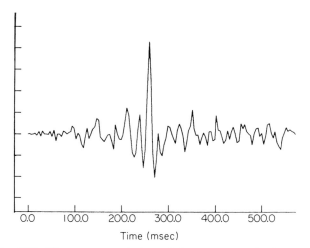

Fig. 2.32. The cross-correlation of the responses of Figs. 2.30 and 2.31. Note the ease with which the peak can be picked.

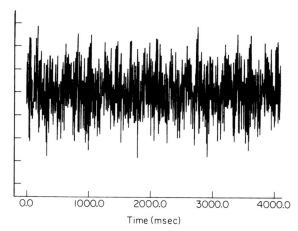

Fig. 2.33. A longer realisation of a Gaussian noise time series of zero mean and unit variance.

The degree of correlation is quantified in practice by using the normalised cross-correlation function. This is defined for real time series as

$$g'_k(x) = \frac{\sum_{t=0}^{N-1} \bar{x}_t \cdot y_{t+k}}{\left[\sum_{t=0}^{N-1} x_t^2\right]^{1/2} \cdot \left[\sum_{t=0}^{N-1} y_t^2\right]^{1/2}} \qquad (2.7.5)$$

for $k = 0, 1, \ldots, N-1$

The absolute value of this function never exceeds 1. Perfect correlation corresponds to the value 1, perfect anti-correlation corresponds to the value -1 and no correlation whatsoever corresponds to the value 0. This normalised form of the cross-correlation function is used extensively in a number of areas in seismic data processing, including velocity analysis and residual statics estimation. Further discussion is again deferred to Chapter 3.

2.7.2 Noise and its auto-correlation

For a random zero-mean noise series n_t of standard deviation d, the auto-correlation function is defined as

$$r_k(n) = \frac{1}{N} \sum_{t=0}^{N-1} \bar{n}_t \cdot n_{t+k} \qquad k = 0, 1, \ldots, N-1 \qquad (2.7.6)$$

Using arguments similar to those above,

$$E(n_t \cdot n_{t+s}) = d^2 \qquad \text{for } t = s$$
$$= 0 \qquad \text{otherwise}$$

Furthermore, it can be shown (Kendall and Stuart, 1966), that

$$E(r_k(n)) = -\frac{1}{N} \qquad (k > 0) \qquad (2.7.7)$$

with a variance of $\frac{1}{N}$

By way of example, Fig. 2.33 shows a longer realisation of the random noise shown in Fig. 2.27. The auto-

correlation of this is shown in Fig. 2.34. Note that this is actually a good way of detecting random noise as its auto-correlation should behave roughly as indicated by (2.7.7). Any big departures from this behaviour are indicative of some non-random effect.

Finally, a most important property of random numbers is that the sum of N' random numbers of average magnitude 1 adds up to

$$(N')^{1/2} \qquad (2.7.8)$$

This effect is widely exploited in seismic data processing, particularly in the process known as stacking (see Chapter 3).

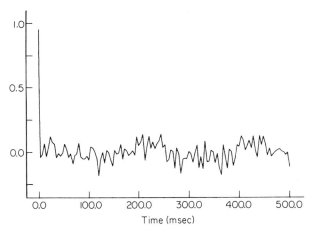

Fig. 2.34. The auto-correlation of the noise series of Fig. 2.33.

2.7.3 Periodicities in correlation functions

It was stated in Section 2.7.1 that one of the important properties of the auto-correlation function was that if the input function was periodic with period T, then the auto-correlation was also periodic with period T. This will now be proved, again for the discrete case.

Suppose the input time series is periodic with period N samples and is given by

$$x = \cos\left(\frac{2\pi l}{N}\right) \qquad l = 0, 1, \ldots, N-1 \qquad (2.7.9)$$

Then the auto-correlation is given by

$$r_k = \frac{1}{M} \sum \cos\left(\frac{2\pi l}{N}\right) \cdot \cos\left(\frac{2\pi(k+l)}{N}\right) \quad (2.7.10)$$

$$= \frac{1}{2M} \sum_{l=0}^{M-1} \left[\cos\left(\frac{2\pi(k+2l)}{N}\right) - \cos\left(\frac{2\pi k}{N}\right)\right]$$

$$(2.7.11)$$

This is the difference of two geometric series which after some manipulation gives

$$r_k = O\left(\frac{1}{M}\right) + \frac{1}{M} \cdot M \cdot \cos\left(\frac{2\pi k}{N}\right) \quad (2.7.12)$$

It may then be seen, as M tends to infinity, the discrete auto-correlation function tends to a periodic function of the same period as the input.

Using a similar argument to that above, and starting with a function

$$x_l = \cos\left(\frac{2\pi l}{N} + P\right)$$

the following closely related and equally important result can be proved: the auto-correlation function retains none of the phase information of the input signal.

Hence, the auto-correlation of a minimum-phase wavelet is exactly the same as the auto-correlation of any other wavelet of the same amplitude spectrum. On the other hand, the cross-correlation function does retain information about the phase of the input functions.

2.7.4 Correlations and signal-to-noise ratio

Signal-to-noise ratio is a particularly abused concept, not only in exploration seismology, but also in other sciences. Heuristically, it is an attempt to compare the 'size' of a signal embedded in noise with the 'size' of that noise. The problem is that there is no unique definition and so it is impossible to speak of the signal-to-noise ratio. In theoretical textbooks, it is often defined in terms of a signal of infinite duration embedded in noise of a similar duration. In seismology, however, the nature of the seismogram implies that signal is actually only present in certain discrete zones corresponding to changes in acoustic impedance. Where there is no reflection, there is no signal.

In such circumstances, it makes sense to measure signal-to-noise ratio only over those zones or windows where signal is present. If these windows are short, simple statistics states that the estimator in use may have an intolerably large variance or uncertainty, simply because there are insufficient samples within the window. Suffice it to say that it is not an easy concept to define in this environment.

Some of the possible candidates for estimators over a window in which there is signal are:

$$\frac{\text{peak to peak signal}}{\text{peak to peak noise}} \quad (2.7.13)$$

$$\frac{\text{RMS of the signal}}{\text{RMS of the noise}} \quad (2.7.14)$$

$$\frac{\text{average absolute amplitude of the signal}}{\text{average absolute amplitude of the noise}}$$

$$(2.7.15)$$

These different ratios will differ markedly as the spectrum and distribution of the noise changes. It is often assumed to be Gaussian but this assumption is only approximately true and frequently wildly inaccurate.

The great problem in practice, of course, is that the seismogram contains signal plus noise, whereas signal-to-noise is a ratio. As will now be shown, if certain reasonable assumptions are made, signal-to-noise ratios are intimately related to correlation functions.

Assume that there are at least M seismic traces which contain the same wavelet, perhaps with some relative time shift and different realisations of additive noise. Also, again temporarily, continuous time functions will be assumed with a corresponding change in nomenclature. The general composite trace, wavelet plus noise, can be written

$$c_j(t) = a_j \cdot w(t - h_j) + n_j(t) \quad j = 1, \ldots, M$$
$$(2.7.16)$$

where $w(t)$ is the signal wavelet
  h_j is the time shift on trace j
  a_j is the amplitude of the wavelet on trace j
  $n_j(t)$ is the additive noise on trace j.

For uncorrelated, zero mean noise,

$$E[n_i(t) \cdot n_j(t+h)] = 0, \quad i \neq j$$
$$E[w_i(t) \cdot n_j(t+h)] = 0, \quad \text{all } i \text{ and } j. \quad (2.7.17)$$

Also define the continuous time cross-correlation of traces $c_i(t)$ and $c_j(t)$ as a function of lag h as

$$g_{ij}(h) = E[c_i(t) \cdot c_j(t+h)] \quad (2.7.18)$$

Then using (2.7.16) and (2.7.17)

$$g_{ij}(h) = a_i a_j E[w(t-h_i) \cdot w(t-h_j+h)]$$
$$= a_i a_j r_w[h - (h_j - h_i)] \quad (2.7.19)$$

where $r_w(h)$ is the auto-correlation function of w at lag h.

Similarly, from (2.7.16), the auto-correlation of trace c_i is

$$r_{c_i}(h) = E[c_i(t) \cdot c_i(t+h)]$$
$$= a_i^2 r_w(h) + r_{n_i}(h) \quad (2.7.20)$$

where $r_{n_i}(h)$ is the auto-correlation of the noise at lag h.

Referring to the form of (2.7.5), the continuous time normalised cross-correlation may be defined using (2.7.19) and (2.7.20) as

$$g'_{ij}(h) = \frac{g_{ij}(h)}{[r_{c_i}(0)]^{1/2} \cdot [r_{c_j}(0)]^{1/2}}$$

$$= \frac{a_i a_j r_w[h - (h_j - h_i)]}{[a_i^2 r_w(0) + r_{n_i}(0)]^{1/2} \cdot [a_j^2 r_w(0) + r_{n_j}(0)]^{1/2}}$$

$$(2.7.21)$$

As has already been noted, the auto-correlation has its maximum value when the lag is zero, hence, g'_{ij} is a maximum for $h_M = h_j - h_i$, and the maximum value is given by

$$[g'_{ij}]_M = \frac{a_i a_j r_w(0)}{[a_i^2 r_w(0) + r_{n_i}(0)]^{1/2} \cdot [a_j^2 r_w(0) + r_{n_j}(0)]^{1/2}}$$

$$(2.7.22)$$

This is the desired result but there remains the question of how it can be used. Recall that the object is to relate the cross-correlation to a signal-to-noise ratio, S_N say. Consider the choice (2.7.14). Since the zero-lag value of the auto-correlation is no more than the sum of squared magnitudes, it follows that

$$[S_N]_i = \frac{\text{RMS of signal}}{\text{RMS of noise}}$$

$$= \frac{a_i[r_w(0)]^{1/2}}{[r_{n_i}(0)]^{1/2}}$$

$$(2.7.23)$$

for trace i. Combining (2.7.22) and (2.7.23) gives

$$[g'_{ij}]_M^2 = \frac{1}{\left[1 + \frac{1}{[S_N]_i^2}\right] \cdot \left[1 + \frac{1}{[S_N]_j^2}\right]}$$

$$(2.7.24)$$

Note that since

$$0 \leqq [S_N]_{i,j} \leqq \text{infinity}$$

then

$$0 \leqq |g'_{ij}| \leqq 1$$

as before.

As it stands, (2.7.24) gives one equation and two unknowns, the two signal-to-noise ratios. There are two ways around this. First, an additional trace c_k could be brought in, giving three equations in three unknowns. If more traces are brought in, statistical redundancy can be used to improve the estimates, remembering that in practice (2.7.24) must be approximated for discrete finite time series with a corresponding statistical uncertainty. Second, simple assumptions could be made to close the loop. For example, if

$$[S_N] = [S_N]_i = [S_N]_j$$

that is, if the signal-to-noise ratios on traces i and j are assumed identical, then

$$[S_N] = \left[\frac{[g'_{ij}]_M}{1 - [g'_{ij}]_M}\right]^{1/2}$$

$$(2.7.25)$$

In the next section, the powerful filtering techniques due to Wiener will be introduced and analysed.

2.8 The Wiener filter

2.8.1 Convolution as a matrix equation

As was seen earlier, the discrete linear convolution of two time series, h and x say, is defined by (cf. equation (2.4.1)),

$$y_k = \sum_{j=0}^{L_h} h_j \cdot x_{k-j}$$

$$k = 0, \ldots, L_s(= L_h + L_x - 1) \qquad (2.8.1)$$

Here L_h is the length of the filter h, L_x is the length of the input x, and L_y is the length of the actual output y.

In this section, it will be shown that (2.8.1) can be written as a matrix equation, thus allowing the power of matrix algebra to be applied to the understanding of this fundamental equation. The appropriate form is

$$\begin{bmatrix} y_0 \\ \cdot \\ \cdot \\ \cdot \\ \cdot \\ \cdot \\ y_{L_y} \end{bmatrix} = \begin{bmatrix} x_0 & 0 & & \ldots & 0 \\ x_1 & x_0 & 0 & \ldots & \cdot \\ x_2 & x_1 & x_0 & \ldots & \cdot \\ x_3 & x_2 & x_1 & \ldots & \cdot \\ \cdot & \cdot & \cdot & \ldots & \cdot \\ \cdot & \cdot & \cdot & \ldots & \cdot \\ 0 & 0 & 0 & & x_{L_x} \end{bmatrix} \begin{bmatrix} h_0 \\ \cdot \\ \cdot \\ \cdot \\ \cdot \\ h_{L_h} \end{bmatrix} \qquad (2.8.2)$$

This may easily be verified to be the same as (2.8.1). It can be written as

$$\mathbf{Y} = \mathbf{XH} \qquad (2.8.3)$$

[Note that boldface letters will be used to denote matrix notation throughout this book.]

where $\mathbf{Y}$ is the column vector $[y_0, \ldots, y_{L_y}]^T$
$\mathbf{X}$ is the matrix shown in equation (2.8.2)
$\mathbf{H}$ is the column vector $[h_0, \ldots, h_{L_h}]$.

(2.8.2) and (2.8.3) are also known as the complete transient convolution of x with h.

2.8.2 The method of least squares

This technique is extremely common throughout exploration seismology, just as in other sciences, and dates back to Gauss. The first principles can be found in many textbooks, and only the following problem will be considered here: suppose a filter h is required which, when convolved with a given input x, produces a desired output d.

In practice, for discrete finite time series, there is no such filter in general and the question must be refined to be what is the filter which gets closest to producing the desired effect in some sense.

The least-squares solution to our problem may be stated as that filter which minimises

$$I = \sum_{k=0}^{L_h+L_x} (d_k - y_k)^2 \qquad (2.8.4)$$

i.e. the sum of squared differences of desired and actual outputs. This least-squares minimisation criterion is also known as minimising using the L_2 norm, taking care not to confuse this with the use of L as a length as above. Note that there are an infinite number of other ways of minimising the error in some sense. For example, we could choose a filter which minimised the sum of the absolute differences in amplitude

$$I = \sum_{k=0}^{L_h + L_x} |d_k - y_k| \qquad (2.8.5)$$

which corresponds to minimising in the so-called L_1 norm, as considered by Claerbout and Muir (1973). Much current research centres around these other techniques, but we will only consider the problem of minimising (2.8.4).

(2.8.3) and (2.8.4) together give:

$$I = \sum_{k=0}^{L_h + L_x} \left(d_k - \sum_{t=0}^{L_h} h_t \cdot x_{k-t} \right)^2 \qquad (2.8.6)$$

Now expand our definitions of x, d, and h to re-introduce the concept of time-zero as follows:
Let x_l where $\qquad l = -P, \ldots, 0, \ldots, Q$
be the input and
let d_m where $\qquad m = -R, \ldots, 0, \ldots, S$
be the desired output and
let h_n where $\qquad n = -T, \ldots, 0, \ldots, U$
be the required filter, which is currently unknown. The problem then, is to minimise

$$I = \sum_{k=-P-T}^{U+Q} \left(d_k - \sum_{t=-T}^{U} h_t \cdot x_{k-t} \right)^2 \qquad (2.8.7)$$

by varying the h_t.

Employing the standard technique of taking the partial derivative with respect to h_t, for all t, and setting equal to 0 gives

$$\sum_{k=-P-T}^{U+Q} d_k \cdot x_{k-t} = \sum_{k=-P-T}^{U+Q} \sum_{t=-T}^{U} h_t \cdot x_{k-t} \cdot x_{k-j}$$

for $j = -T, \ldots, 0, \ldots, U$

Interchanging the orders of summation and rewriting yields finally

$$\sum_{k=-P-T}^{U+Q} x_{k-t} \cdot x_{k-j} \sum_{t=-T}^{U} h_t = \sum_{k=-P-T}^{U+Q} d_k \cdot x_{k-t}$$

$$(2.8.8)$$

for $j = -T, \ldots, 0, \ldots, U$

This is the result of interest, and is known as the discrete Wiener–Hopf equation, following the pioneering work of the great American mathematician Norbert Wiener. The first term on the left-hand side of this equation is the auto-correlation which is known, the second term is the filter coefficients, which are unknown, and the right-hand side is the cross-correlation of the input with the desired output, which is also known.

Note for now that (2.8.8) can be written in the matrix formulation introduced earlier as

$$\mathbf{X}^T \mathbf{X} \mathbf{H} = \mathbf{X}^T \mathbf{D} \qquad (2.8.9)$$

(2.8.8) is actually a set of simultaneous equations which have been solved in the order of ten thousand million times during the era of digital computer technology in seismology alone. In spite of its immense practical importance in discrete time series analysis, Wiener's work is surprisingly little known outside the fields of communication theory, seismology and econometrics.

Before continuing with the discussion of (2.8.8) and its many uses, it will be transformed to facilitate its application in a FORTRAN computer program, where the use of negative indices is either forbidden or fraught with danger, depending on the compiler implementation.

Let

z_x be the index of time zero in the input
z_d be the index of time zero in the desired output
z_h be the index of time zero in the filter.

Then

$$z_x = P + 1$$
$$z_d = R + 1 \qquad (2.8.10)$$
$$z_h = T + 1$$

Now define the shifted arrays and indices

$$x'_{l+P+1} = x \qquad l = -P, \ldots, 0, \ldots, Q$$
$$d'_{m+R+1} = d \qquad m = -R, \ldots, 0, \ldots, S$$
$$h'_{n+T+1} = h \qquad n = -T, \ldots, 0, \ldots, U$$
$$j' = j + T + 1 \qquad (2.8.11)$$
$$t' = t + T + 1$$
$$k' = k + P + T + 1$$

Substituting (2.8.10) and (2.8.11) in (2.8.8) gives after some manipulation

$$\sum_{k'=1}^{L_h + L_x - 1} x'_{k'-t'+1} \cdot x'_{k'-j'+1} \sum_{t'=1}^{L_h} h'_{t'}$$
$$= \sum_{k'=1}^{L_h + L_x - 1} d'_{k'+z_d+1-z_x-z_h} x'_{k'-t'+1} \qquad (2.8.12)$$

for $t' = 1, \ldots, L_h$, where L_x and L_h are the total lengths of the input and filter respectively. The reworked version of (2.8.8) is in a form suitable for inclusion in a computer program.

Note that a knowledge of z_x is unnecessary for the calculation of the left-hand side of (2.8.12), another manifestation of the fact that the auto-correlation has no phase information.

2.8.3 The Wiener filter and its uses

(2.8.8) occurs in many guises in seismic data processing and is at the heart of many inverse filtering techniques. In addition, it represents a classic example of inversion as elucidated in Chapter 5. For now, the following common areas of application will be described:

2.8.3.1 Signature shaping

Very often in seismology, it is desirable to replace a known response a, say, with another of a more convenient temporal 'shape'. For example, if it is desired to replace a wavelet by its equivalent zero-phase wavelet, then letting $x = a$ and $d = b$ in (2.8.8), the resulting filter h is the Wiener shaping filter which when convolved with a will turn it into b (approximately, in the least-squares sense). Hence, if this filter is convolved with a seismogram in which a occurs in various places, each occurrence will be turned into b. The effect on the rest of the data, which is convolutionally formed (the basic approximation of the convolutional model) is essentially nothing. This is often written in the following way: Referring to (2.4.3), suppose

$$s = a * e + n$$

Then convolving with the filter h gives

$$h * s = (h * a) * e + h * n$$
$$= b * e + h * n$$

The second term on the right-hand side is just filtered noise.

2.8.3.2 Wiener inversion

In Wiener inversion, the intention is to replace a given wavelet within the data by a single 'spike', and it is necessary to convolve the data with the inverse of the given wavelet. For this, let $x = a$ in (8.8) and $d = 1$.

This is simply a special case of Wiener shaping described above. The ability with which Wiener shaping or inversion can be performed depends on the relationship between the phase of the input and the desired output. If these are noticeably disparate, it may be necessary to design an inverse filter with a combination of memory and anticipation components (coefficients after and before time-zero respectively), in order to perform the shaping effectively. Probably the most prevalent form of Wiener filtering however, is considered in the next section.

2.8.3.3 Predictive deconvolution

In practical terms, this is one of the most important parts of this chapter.

Consider the problem of attempting to predict future values of a time series from past and present values.

Let the input x be some time series recorded up to the present time index, t say, i.e.

$$\ldots, x_{t-3}, x_{t-2}, x_{t-1}, x_t \qquad (2.8.13)$$

Now let the desired output be the input, q time units into the future, i.e.

$$d_t = x_{t+q} \qquad (2.8.14)$$

q is called the 'prediction distance' or 'lag'.

Using (2.8.13) and (2.8.14) in (2.8.8) gives

$$\sum_{k=-P-T}^{U+Q} x_{k-t} \cdot x_{k-j} \sum_{t=-T}^{U} h_t$$
$$= \sum_{k=-P-T}^{U+Q} x_{k+q} \cdot x_{k-t} \qquad (2.8.15)$$

for $j = -T, \ldots, 0, \ldots, U$

Suppose the solution of (2.8.15) in matrix notation is the prediction filter

$$\mathbf{h}(d)$$

This filter, when applied to the time series x, attempts to predict x_{t+q} from the input x_t. The success of this operation will obviously depend on how predictable the time series x_t is.

Where is all this leading? The important point in seismology is that multiples are highly predictable generally, whereas the reflection series, with which they are convolved, is normally highly unpredictable (this is completely untrue for cyclic depositional sequences). Hence, predictive deconvolution, if used correctly, ought to be a very useful tool for the attenuation of multiple energy, particularly of a short period. This generally turns out to be the case although the assumptions will be scrutinised much more closely in Chapter 3.

The predictive deconvolution filter is constructed as follows. Consider the filter

$$(1, \underbrace{0, 0, 0, \ldots, 0}_{q-1 \text{ zeroes}}, -\mathbf{h}(d)) \qquad (2.8.16)$$

formed from the filter $\mathbf{h}(d)$, which is the solution of (2.8.12). If this filter is convolved with the data, then that part of the data q units into the future, which is predictable from past and present values, should be predicted and subtracted out by the application of (2.8.13).

(2.8.13) is usually known as the prediction error filter. The prediction distance q is chosen to be the period of the multiple energy it is desired to remove, as illustrated by Fig. 2.35. With the comments of Section 2.7 in mind,

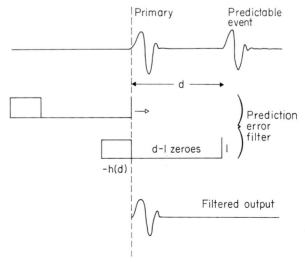

Fig. 2.35. A schematic of the action of predictive deconvolution. The prediction distance or lag is d samples.

the prediction distance or lag, q, is often picked from the auto-correlation.

2.8.4 White light

Two other parameters are at the disposal of the geophysicist in the design of the predictive deconvolution filter, apart from the lag. The first of these is the filter length and more will be said of this in Chapter 3. The second is somewhat more esoteric. This is the amount of white light which must be included in the Wiener filter design. Mathematically, this corresponds to the addition of a constant percentage of the zero-lag value of the auto-correlation to each element of the main diagonal of the matrix equation (2.8.12), which is simply the zero-lag value of the auto-correlation repeated.

Using (2.8.9), this may be written

$$(\mathbf{X}^T\mathbf{X}\mathbf{H} + l_w\mathbf{I}) = \mathbf{X}^T\mathbf{D} \qquad (2.8.17)$$

where l_w is the amount of white light, assuming that the auto-correlation has been normalised to have a zero-lag of 1, and $\mathbf{I}$ is the identity matrix. This performs two tasks:

1 It stabilises the filter calculation, because the addition of a constant to the diagonal has the effect of adding a small constant to all eigenvalues of the matrix, hence rendering zero eigenvalues non-zero. The amount required to achieve this depends on the word length of the computer and the length of the filter. Typically, around 0.0001 per cent is quite sufficient in everyday usage.

2 Most importantly for the geophysicist, it controls the gain, i.e. amplification at those frequencies which are poorly represented in the input data. To see this, remember that the filter is applied convolutionally. Consider the action in the frequency domain. The frequency domain equivalent to equation (2.8.3) is

$$H \cdot X = Y \qquad (2.8.18)$$

where H is the Fourier transform of h
X is the Fourier transform of x
Y is the Fourier transform of Y and any scalar factors which have been absorbed.

Hence, if frequencies present in the output Y, are poorly represented in the input X, the multiplicative effect of H, or gain, must be correspondingly large. Since poor representation is usually (but not always) indicative of poor signal-to-noise ratio, the judicious usage of white light can prevent undesirable amplification of noise in the data as the equivalent of (2.8.18) is then

$$H \cdot (X + l) = Y \qquad (2.8.19)$$

where l is proportional to the amount of white light and is independent of frequency. Where X is large, the effect of l is small, but where X is small, the presence of l controls the gain.

There is an interesting relationship between the lag, the amount of white light and the amplification of unwanted noise on real seismic data which it generally

takes considerable experience to acquire. Simplistically however, the amplification decreases with increasing white light and decreases with increasing lag. For gain control, much larger amounts of white light are used, typically of the order of 0.5–5.0 per cent.

For a more detailed treatment of this topic, see Section 5.8.

Finally, it is worth noting that the Toeplitz structure is a feature of equation (2.8.9). This symmetry is exploited by a fast algorithm for solving such equations known as the Levinson algorithm after its inventor. The details of this important algorithm can be found in Robinson and Treitel (1980).

2.9 Spectral analysis

2.9.1 Introduction

This topic will be considered from a rather individual viewpoint, treating the whole subject as the accumulation of largely convolutional effects in the frequency domain.

The conventional approach to spectral analysis whereby a time series is decomposed into its Fourier components, was originally based around a computation of the auto-correlation of the time series whose spectrum was desired. The relevance of this can be seen after the following important result is derived.

First note that an estimate of the power spectrum of the N-sample time series $x_0, x_1, x_2, \ldots, x_{N-1}$ may be obtained by

$$\hat{P}(z) = \frac{1}{N}\,\bar{X}\left(\frac{1}{z}\right) \cdot X(z) \qquad (2.9.1)$$

where $\hat{P}(z)$ is the power spectral estimate and $X(z)$ is the Z-transform of x. Now by definition, the power spectrum is the expected value of $\hat{P}(z)$, i.e.

$$P(z) = E[\hat{P}(z)]$$

As can easily be verified, substituting the auto-correlation formula (2.7.1) into (2.9.1) and equating like powers of z gives the important result that the power spectrum is the Fourier cosine transform of the auto-correlation function.

Using this result, the power spectrum could then be computed directly from the auto-correlation. This technique was originally found desirable for two main reasons. First, only a few lags of the auto-correlation were generally required to define it well and so the resulting transform was quite short, enabling it to be calculated without too much computation. Second, as we have seen, the auto-correlation is much less affected by the presence of noise.

The advent of the FFT made it possible to do a direct Fourier transform of the input time series, and then smooth the transform using various smoothing filters to reduce the effects of noise (i.e. to stabilise the estimate). It is generally this 'tinkering' which has to be done to cope with the effects of noise which makes the subject a little more difficult to comprehend.

2.9.2 A convolutional model of the spectrum

Leaving aside the question of contaminating noise for the moment, the basic problem may be illustrated as follows.

Let $s(t)$ be a continuous function of time and let $S(f)$ be its Fourier transform, so

$$S(f) = s(t) \cdot \int_{-\infty}^{\infty} \exp\left(-i(2\pi f t)\right) \cdot dt \qquad (2.9.2)$$

and define the following two functions

$$\text{rect}\left(\frac{t}{T}\right) = 1 \qquad |t| \leq \frac{T}{2}$$
$$= 0 \qquad \text{otherwise} \qquad (2.9.3)$$

which has the transform $T \dfrac{\sin(Tf)}{Tf}$ or $T \, \text{sinc}(Tf)$,

$$\text{shah}\left(\frac{t}{d}\right) = 1 \qquad \frac{t}{d} \text{ an integer}$$
$$= 0 \qquad \text{otherwise} \qquad (2.9.4)$$

which has the transform $d \, \text{shah}\,(fd)$ (see Bracewell, 1978).

The argument will now be developed in both time and frequency domains.

Time	Frequency
$s(t)$	$S(f)$

Sample in time by multiplication by the shah function $\text{shah}(t/d)$, where d is the sampling interval.

Time	Frequency	
$\text{shah}(t/d) \cdot s(t)$	$d \cdot \text{shah}(tf) * S(f)$	(2.9.5)

Take a finite window of duration T by multiplication by a rectangle function.

Time
$$\text{rect}(t/T) \cdot \text{shah}(t/d) \cdot s(t)$$

Frequency
$$Td \cdot \text{sinc}(ft) * \text{shah}(ft) * S(f) \qquad (2.9.6)$$

Note that it is immaterial what $s(t)$ consists of (i.e. signal, noise, or signal plus noise). The right-hand part of expression (2.9.6) exactly expresses what happens to the original spectrum $S(f)$ by the physical act of sampling and windowing as expressed by the left-hand part of expression (2.9.6).

The central problem of spectral analysis is to recover $S(f)$ from

$$Td \cdot \text{sinc}(ft) * \text{shah}(ft) * S(f) \qquad (2.9.7)$$

in a way which is stable in the presence of noise. Fig. 2.36 illustrates this convolutional build-up of perturbing effects in the frequency domain.

What are the effects of this spectral distortion?
1 The effect of $\text{sinc}(fT)$ is to 'blur' or 'smear' $S(f)$, with the result that close spectral lines in $S(f)$ may coalesce and become invisible.

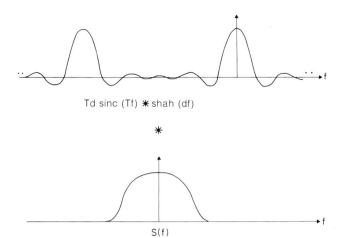

Fig. 2.36. A frequency domain representation of the convolutional build-up of windowing and aliassing effects.

2 The effect of $\text{shah}(fd)$ is to duplicate the spectrum with a periodicity of $1/d$. This leads to the phenomenon of aliassing as was discussed in Section 2.3.3.

As far as the smearing is concerned, it should be noted that it is at its worst when the input spectrum is changing rapidly. In practice, attempts are made to compensate for this by removal of obvious 'trends', including the mean (D.C.) and linear trend, and in some cases more complicated functions, before the spectral analysis is performed. These techniques are known generically as pre-whitening. After the spectral analysis has been done, the removed effects can be re-incorporated if desired.

Pre-whitening can be done prior to all of the spectral analysis techniques discussed below.

2.9.3 The transient sampling theorem

Using the same convolutional model introduced above, Hatton (1981) showed that the effects of 'smearing' and aliassing cannot be separated when the number of distinct samples $T/d = N$, say, is small. In that event, the case of no-aliassing is given by a theorem which he referred to as the 'transient sampling theorem'. This theorem states that if d is the sampling interval which just avoids aliasing in a signal of infinite duration, as given by the Nyquist criterion discussed earlier, then the sampling interval d' necessary to avoid aliasing of an N-sample subset of this signal is given by

$$d' = \frac{d}{1 + \dfrac{1}{2N}} \qquad (2.9.8)$$

The phenomenon arises because of the finite width of the 'windowing' function, $\text{sinc}(fT)$ in this case.

Returning to the problem of extracting the true spectrum and assuming that d' is chosen sufficiently small to avoid the effects of (2.9.8), the problem still remains of how to remove the effects of smearing as noted in **1**. The classical method of getting round this will now be compared with two newer methods. First the classical method will be considered.

2.9.4 Data-independent spectral analysis (window carpentry)

The essence of window carpentry is to find a time domain function other than rect(t/T) with which to window the data, which has better frequency domain properties in some sense. The two relevant properties are:

1 The width W of the main 'lobe' of the transform of the windowing function.

2 The relative sizes of the main 'lobe' B, compared to the side 'lobes' A, of the transform of the windowing function.

Fig. 2.37 illustrates the nomenclature.

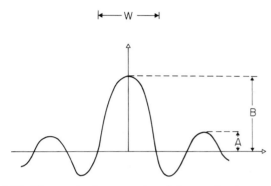

Fig. 2.37. The two basic properties of the amplitude frequency response of a spectral windowing function. W is the width of the main lobe. B/A is the ratio of the amplitudes of the main lobe and the biggest side lobe.

It turns out to be impossible to reduce both W and A/B simultaneously. A compromise must be made. This compromise is at the heart of window carpentry. There are many such choices of which two of the most well-known ones are the windowing functions due to Bartlett and Parzen as described, for example, by Kanasewich (1981).

The reason why window carpentry is called data-independent is that it involves modification of the window function only, whatever the data within the window.

2.9.5 Data-adaptive spectral analysis (MLM and MEM)

In the late 1960s and early '70s two radically new methods of spectral analysis appeared. Both are data-dependent in the sense that they modify their spectral smearing according to the data contained within the window. This is fundamentally different to window carpentry, and the methods are capable of giving much higher resolution for short time series, although they give essentially the same results for longer time series.

The MLM or maximum likelihood method due to Capon (1969), involves a window which is designed to reject all frequency components in an optimal way except for the one frequency component which is desired. The optimality criterion is derived by demanding that a pure sinusoid without noise be passed unaf-

fected, whilst a pure noise input is minimised. Again, the reader is referred to Kanasewich (*loc. cit.*) for algorithmic details.

The MEM or maximum entropy method due to Burg (1967), is particularly abstract in that it involves the design of a prediction error filter which optimally 'whitens' the data. This corresponds to maximising the amount of information in the input. In effect, the required spectrum is the inverse to that of this whitening filter. It may be thought of as predicting the given time series beyond the bounds of the window available using optimal prediction theory, thus extending the possible spectral resolution available. Again, see Kanasewich (*loc. cit.*) for details.

It may interest the reader to see a comparison of the various techniques described above. Figs. 2.38–2.41 are a recreation of the example used by Lacoss (1971), which consists of two sinusoids of 0.3 and 0.4 Hz in an approximately 10 per cent white noise background sampled for 10 seconds with a 1 second sample interval. Note that the triangular shapes of the peaks are due to linear interpolation in the plotting at the frequency sample interval used. Fig. 2.38 is the resulting power spectrum and Figs. 2.39–2.41 show the Bartlett, MLM and MEM estimates respectively. The Bartlett estimate is incapable of separating the two close peaks. The MLM method is just capable of resolving them whereas

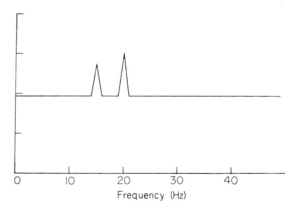

Fig. 2.38. An idealised power spectrum representing two sinusoids buried in white noise. The lower amplitude spike is 6 db down on the larger.

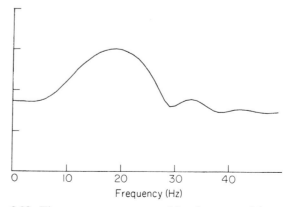

Fig. 2.39. The power spectrum resulting from use of the Bartlett spectral window on the spectrum of Fig. 2.38. Note that the two sinusoids cannot be resolved.

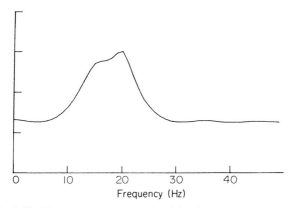

Fig. 2.40. The power spectrum resulting from use of the maximum likelihood method on the spectrum of Fig. 2.38. The two sinusoids can just be resolved. Their amplitude relationship is correctly deduced.

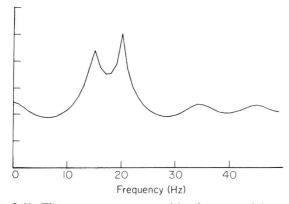

Fig. 2.41. The power spectrum resulting from use of the maximum entropy method on the spectrum of Fig. 2.38. The two sinusoids can easily be resolved. Note that their amplitude relationship is proportional to the area under the peaks in this method, rather than their height.

the MEM method easily resolves them. One odd thing which the user ought to note about the MEM estimate, is that the power is proportional to the area under the peak instead of the size of the peak as is the case with the other methods. This can make MEM estimates a little difficult to interpret by eye.

It is also interesting to note the degradation in all these methods when the background noise is not white. Figs. 2.42–2.45 illustrate for exponentially distributed background noise and Bartlett, MLM and MEM estimates respectively. This degradation is due to convolutional smearing of this non-white background into the two peaks.

In the next section, in order to illustrate some of the above concepts, an important filtering operation in seismology will be spectrally analysed in the simplest possible way, using the direct method.

2.9.6 Example: the spectrum of an interpolator

As was mentioned in Section 2.3.6, the ideal interpolator has a flat spectrum of unit amplitude between D.C. and the Nyquist frequencies, and can only be achieved in practice in special circumstances. In general, interpolators are not ideal and it is of considerable

interest to the processing geophysicist to understand exactly what they do to the spectrum of the data in view of their frequent usage in such processes as normal move-out correction, as discussed further in the chapter on processing.

Since interpolation is equivalent to the convolution of the data with some interpolating filter, a particularly convenient way of assessing the quality of an interpolator is simply to consider its coefficients as a time

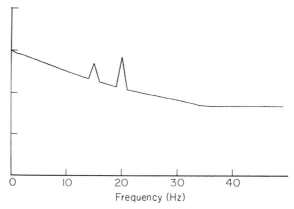

Fig. 2.42. An idealised power spectrum representing two sinusoids buried in Gaussian noise. The lower amplitude spike is 6 db down on the larger.

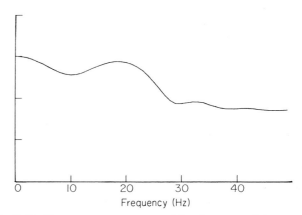

Fig. 2.43. The power spectrum resulting from use of the Bartlett spectral window on the spectrum of Fig. 2.42. Note that the two sinusoids cannot be resolved.

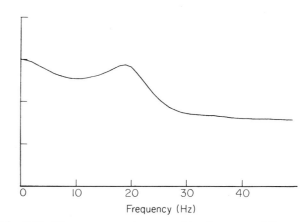

Fig. 2.44. The power spectrum resulting from use of the maximum likelihood method on the spectrum of Fig. 2.42. The two sinusoids cannot now be resolved.

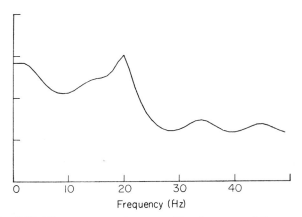

Fig. 2.45. The power spectrum resulting from use of the maximum entropy method on the spectrum of Fig. 2.42. The two sinusoids can just barely be resolved.

series and then to Fourier transform them to inspect how far the transform differs from the ideal case described above.

Doing this for the well-known case of linear interpolation for the mid-point between two known samples, yields an amplitude spectrum shown in Fig. 2.46. It can be seen that the linear interpolator ceases to become accurate much beyond 1/3 of the Nyquist frequency.

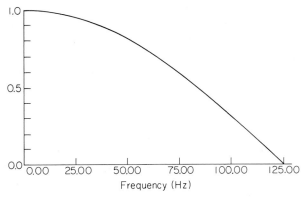

Fig. 2.46. The amplitude spectrum of a linear interpolator for the mid-point of two samples. The spectrum causes unacceptable attenuation beyond about 1/3 of the Nyquist.

Hence use of linear interpolation will severely restrict the bandwidth of data which has frequencies beyond this amount. Linear interpolation is not consequently recommended for seismic data processing unless the bandwidth is already thus limited. Other more complicated interpolators must then be considered. Fig. 2.47 shows the equivalent spectrum of a superior interpolator, the truncated sinc function. In this case, the sinc function consists of 8 points. The amplitude spectrum of this interpolator is much superior to that of the linear interpolator, extending to perhaps 2/3 of the Nyquist frequency. This is entirely adequate for seismic data as this normally represents about the maximum frequency unaffected by the anti-alias filters.

Although not shown here, the phase response of an interpolator is also of importance as it is undesirable to disturb the existing phase response. This is one of the reasons for the attraction of the sinc function.

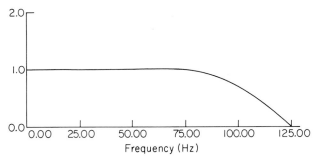

Fig. 2.47. The amplitude spectrum of an 8 point truncated sinc function interpolator for the mid-point of two samples. This spectrum is much broader than the equivalent for a linear interpolator and performs acceptably to about 3/4 of the Nyquist.

The content of this chapter has been concerned with the one-dimensional Fourier transform. In seismic data processing, two-dimensional Fourier transforms are also used extensively. The addition of the extra dimension considerably enriches the properties of the Fourier transform as will now be seen.

2.10 The 2-D Fourier transform

2.10.1 Introduction

The two-dimensional version of the discrete Fourier transform is no more than the one-dimensional transform as expressed by equations (2.3.7) and (2.3.8) extended to two independent variables. These might be two spatial variables, or much more commonly, one temporal variable and one spatial variable. For the moment, the nature of the independent variables is irrelevant and the resulting series will still be referred to as a 'time' series.

Hence, suppose that s_{qr} is a two-dimensional time series defined as

$$s_{qr} \qquad q = 0, 1, \ldots, M-1$$
$$r = 0, 1, \ldots, N-1$$

then its two-dimensional DFT is defined as

$$S_{fk} = \frac{1}{MN} \sum_{q=0}^{M-1} \sum_{r=0}^{N-1} s_{qr}$$
$$\cdot \exp\left(-i2\pi\left(\frac{fq}{M} - \frac{kr}{N}\right)\right) \qquad (2.10.1)$$

where $f = 0, 1, \ldots, M-1$
$k = 0, 1, \ldots, N-1$

and the inverse transform is defined as

$$s_{qr} = \frac{1}{MN} \sum_{f=0}^{M-1} \sum_{k=0}^{N-1} S_{fk} \cdot \exp\left(i2\pi\left(\frac{fq}{M} - \frac{kr}{N}\right)\right)$$
$$(2.10.2)$$

where $q = 0, 1, \ldots, M-1$
$r = 0, 1, \ldots, N-1$

Simple analogues for all of the relationships discussed in Section 2.3.2 exist. For example, periodicity in the transform is expressed by

$$S_{M+f,k} = S_{f,k}$$
$$S_{f,N+k} = S_{f,k}$$

(2.10.3)

The analogues of equations (2.3.10) to (2.3.13) are left to the reader.

On a practical note, equation (2.10.1) can be written as

$$S_{fk} = \frac{1}{M} \sum_{q=0}^{M-1} \exp\left(-i\left(\frac{2\pi fq}{M}\right)\right)$$
$$\cdot \left[\frac{1}{N} \sum_{r=0}^{N-1} s_{qr} \cdot \exp\left(-i\left(\frac{2\pi kr}{N}\right)\right)\right]$$

(2.10.4)

Hence, considering s_{qr} as a matrix, the two-dimensional DFT can be computed by performing a one-dimensional transform on all the rows, followed by another one-dimensional transform on all the columns. A similar treatment shows that the same end is achieved by doing the columns first.

2.10.2 A pictorial study of the 2-D DFT

So far, the tone of this chapter has been set for the reasonably informed physical scientist. Hence, the treatment of the one-dimensional transform concerned itself more with relationships and usage than a pictorial study of the transform itself. An excellent reference for this is Evenden, Stone and Anstey (1971). In the case of the two-dimensional transform, some effort will be expended here to give the reader a more intuitive feel as it represents a data processing tool of great power and flexibility.

In seismology, two-dimensional transforms involving one temporal axis and one spatial axis are by far the most common, and the corresponding two-dimensional frequency domain is usually known as the $f-k$ domain, f and k being the equivalent frequency domain variables to the time and space variables respectively.

To simplify the following discussion, first let $M = N$ and, without loss of generality, re-write equation (2.10.2) using only the cosine parts of the summation as

$$s''_{qr} = \frac{1}{MN} \sum_{f=0}^{M-1} \sum_{k=0}^{N-1} S''_{fk} \cdot \cos\left(\frac{2\pi}{N}(fq - kr)\right)$$

(2.10.5)

where $q = 0, 1, \ldots, M-1$
$r = 0, 1, \ldots, N-1$

Further, consider values of S''_{fk} as shown in Fig. 2.48. There are 10 non-zero values corresponding to $f = 0, 1, \ldots, 9$ Hz, with k chosen so that the values are co-linear in $f-k$. The amplitudes of the S''_{fk} are 1, 2, 2, 2, 2, 2, 1.6, 1.2, 0.8 and 0.4 respectively. The individual terms of the corresponding summation (2.10.5) are shown in Fig. 2.49a–j. When combined together according to equation (2.10.5), Fig. 2.50 results.

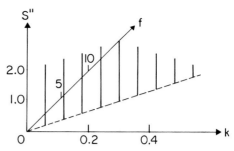

Fig. 2.48. The two-dimensional amplitude spectrum corresponding to 10 two-dimensional sinusoidal functions.

This figure embodies an empirical proof of a very important result, namely that straight lines in $t-x$ transform to straight lines in $f-k$.

A more formal proof will now be presented in some detail and the relationships between corresponding dips in the time domain explored.

2.10.3 Filtering in two dimensions

Filtering in two dimensions is a simple extension of the one-dimensional filtering already considered in Section 2.5. For example, as can easily be shown, two-dimensional convolution is equivalent to multiplication of the two-dimensional transforms. Hence, filtering in two dimensions can be achieved by designing a suitable filter response in the two-dimensional frequency domain and multiplying it with the two-dimensional transform of the data to be filtered, bearing in mind all of the previous caveats about the steepness of the slopes as discussed in Section 2.5.

The attraction of filtering in the $f-k$ domain is that whereas in one dimension, signal and noise may overlap rendering one-dimensional multiplicative filtering impossible, in two dimensions they do not generally overlap in both f and k simultaneously. An example is shown in Fig. 2.51. Here signal and noise overlap in f but not in $f-k$, hence a two-dimensional $f-k$ filter could be designed to reject the noise without affecting the signal. The types of noise and their appearance in the $f-k$ domain will be discussed in detail later but the method is of quite general applicability.

In order to make use of this, the relationship between the two domains must be derived.

Consider the following two-dimensional function

$$s_{qr} = 1 \qquad aq = br + c$$
$$= 0 \qquad \text{otherwise}$$

(2.10.6)

where a, b and c are constants corresponding to a straight line in the $t-x$ domain of slope b/a.

Substituting in (2.10.1) and manipulating eventually gives

$$S_{fk} = \frac{1}{MN} \cdot \exp\left(-i\left(\frac{(M-1)v'q}{2}\right)\right)$$
$$\cdot \frac{\sin(Mv'q)}{\sin(v'q)} \cdot \exp\left(i\left(\frac{2kc}{bN}\right)\right)$$

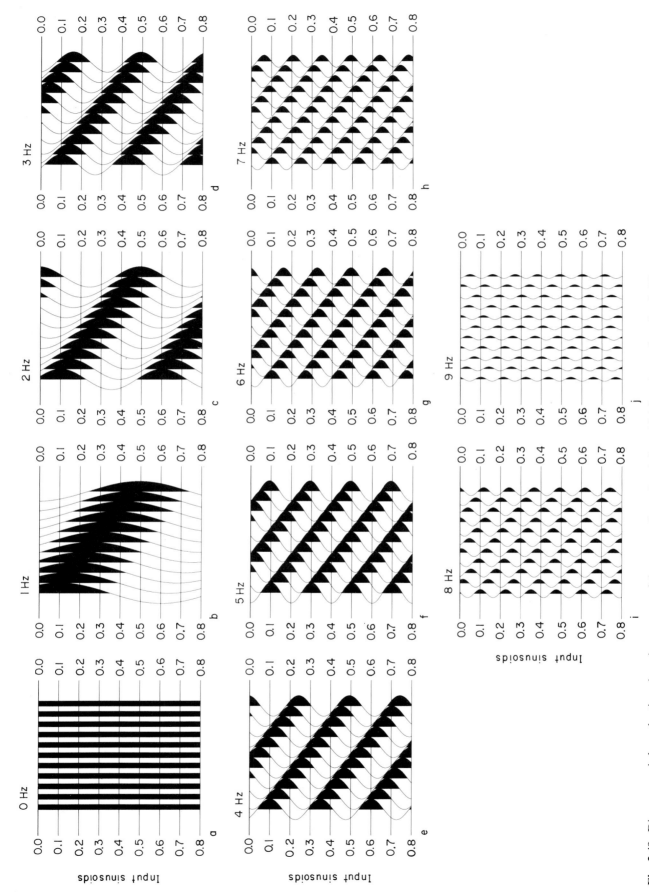

Fig. 2.49. Diagrams a–j show the time domain response of the 10 two-dimensional sinusoidal functions referred to in Fig. 2.48.

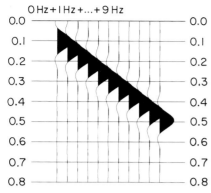

Fig. 2.50. The time domain response corresponding to the two-dimensional frequency domain response of Fig. 2.48.

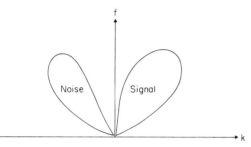

Fig. 2.51. A schematic of the two-dimensional amplitude spectrum illustrating the separation of signal and noise.

where

$$v' = 2\pi\left[\frac{ka}{bN} + \frac{f}{M}\right]$$

Now, letting $M = N \rightarrow \infty$,

$$|S_{fk}| \sim \frac{1}{N} \cdot \left| \frac{\sin\left(2\pi q\left[\frac{ka}{b} + f\right]\right)}{2\pi q\left[\frac{ka}{b} + f\right]} \right| \qquad (2.10.7)$$

In the limit, this corresponds to all energy being concentrated on the line

$$\frac{ka}{b} + f = 0 \qquad (2.10.8)$$

This is the desired result. Note that c appears only as a phase factor.

Fig. 2.52 shows some straight lines and their amplitude spectral $f - k$ transforms, in the way in which they are normally displayed in seismology with f varying

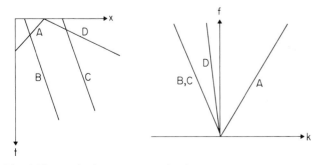

Fig. 2.52. Equivalent responses in the two-dimensional $t - x$ and $f - k$ domains.

from D.C. to the temporal Nyquist f_N, say, and k varying between plus and minus the spatial Nyquist, k_N. Note that because c appears only in the phase of the $f - k$ transform, straight lines B and C map onto the same line in the $f - k$ domain. All such lines must go through the origin in the $f - k$ domain. Furthermore, steepness in the $t - x$ domain, i.e. large values of dt/dx corresponds to shallowness in the $f - k$ domain, i.e. small values of df/dk. Quantifying this, equations (2.10.6) and (2.10.8) show that a line with gradient

$$q = \frac{b}{a} \cdot r + \frac{c}{a} \qquad (2.10.9)$$

in the $q - r$ domain (equivalently the $t - x$ domain), transforms to a line with gradient

$$f = -\frac{a}{b} \cdot k \qquad (2.10.10)$$

in the $f - k$ domain.

Equation (2.10.9) describes a time-distance relationship in which a velocity v_p, may be defined as

$$v_p = \frac{a}{b}$$

v_p is none other than the phase velocity.

Hence (2.10.9) and (2.10.10) can be written as

$$q = \frac{1}{v_p} \cdot r + \frac{c}{a} \qquad (2.10.11)$$

$$f = -v_p \cdot k \qquad (2.10.12)$$

One final note on nomenclature. Since a dipping straight line in $t - x$ transforms to a dipping straight line in $f - k$, events having certain dips between two values in the $t - x$ domain can be removed by multiplying the $f - k$ transform of the data with a transform which is zero between the corresponding dips in the $f - k$ domain and one elsewhere. This is known variously as 'dip-', 'fan-' or 'pie-slice filtering'.

2.10.4 Aliassing and de-aliassing in one and two dimensions

It has already been stated earlier that the extra dimension adds a new richness to the Fourier transform. This is particularly true in terms of the interpretation of aliassing. As has already been discussed for the one-dimensional case in Sections 2.3 and 2.9, aliassing occurs by virtue of discrete sampling in the time domain. This creates duplicates of the input spectrum at multiples of $1/d$ where d is the sampling interval. If the duplicates overlap, the data is said to be aliassed. This nomenclature is somewhat sloppy as aliassing in the strict sense of producing duplicates, occurs by definition in all sampled data, whether overlap occurs or not. If overlapping does not occur, it is a simple job to construct a multiplicative filter in the frequency domain which removes the aliasses or copies of the input spectrum.

In one dimension, if overlapping does occur, its effects cannot be removed without additional information about the sampled spectrum from sources other than the sampled spectrum itself. A trivial example would be if it was known that a 4 millisecond dataset contained frequencies between 0 and 90 Hz along with an unknown amount of aliased 150 Hz. Since 150 Hz produces an alias at $125 - (150 - 125)$ Hz = 100 Hz, the correct spectrum for the 4 millisecond sampling interval could be reconstructed by zeroing out the 100 Hz contribution. Other techniques can be exploited, for example, uneven sampling in the time domain to produce an equivalent of equation (2.9.5) which can be convolutionally inverted (the shah function appearing there is highly singular).

In two and more dimensions, things are much different. For the sake of clarity, the subsequent argument will be presented in terms of the continuous two-dimensional Fourier transform. The argument will be developed analogously to that of Section 2.9 except that an infinite time series will be assumed to avoid the complications due to windowing which led to the Transient Sampling Theorem. Suppose that $s(t, x)$ is a continuous two-dimensional function with a Fourier transform $S(f, k)$. The two-dimensional sampled form of this is given by

$$\text{shah}(t/d) \cdot \text{shah}(x/g) \cdot s(t, x) \qquad (2.10.13)$$

where the shah function was defined in Section 2.9 and d and g are sampling intervals in the time domain and space domain respectively. In the frequency domain, the equivalent of equation (2.10.13) is

$$dg \cdot \text{shah}(fd) * \text{shah}(kg) * S(f, k) \qquad (2.10.14)$$

If it is further supposed that $s(t, x)$ is a straight line in the $t - x$ domain, then as has already been seen, $S(f, k)$ is also a straight line in the $f - k$ domain. Fig. 2.53

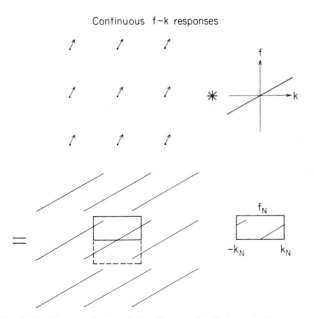

Fig. 2.53. The synthesis of the discrete $f - k$ domain by two-dimensional convolution of the sampling function and the continuous $f - k$ domain response.

illustrates the production of aliasses by the shah $*$ shah function in equation (2.10.8). Note that these aliasses do not overlap in the $f - k$ domain. It is then perfectly possible to extract the original spectrum by suitable multiplicative filtering. In the f or k domains alone, the aliasses do overlap and so the aliassing effect is not removable in a one-dimensional treatment.

This admits of a method of recording a one-dimensional signal in fully recoverable form using a sample interval which would normally alias the data. If the one-dimensional signal is fed into several parallel channels, with a linear analogue delay dependent on channel number applied, the resulting signal could then be discretised and the results displayed as a $f - k$ spectrum. Because of the linear delay, the aliasses will not overlap and so the original spectrum can be fully recovered within the bandwidth applicable to the temporal sample interval. Of course, it could be argued that the same effect could be achieved on one channel simply by increasing the sampling rate. The above technique however could be useful when this is not possible, for example, if the sampling rates required are too high for the available equipment.

2.10.5 Spatial aliassing in seismic data

In this section, the topic of spatial aliassing will be considered. Spatial aliassing is one of those concepts which often arises in conversation about seismic data, but is frequently not well understood at the intuitive level.

Spatial aliassing is defined to have occurred whenever wrap-round occurs in the $f - k$ domain. The mathematical criterion for this is discussed in Section 4.6. Fig. 2.54 illustrates. Event A is not spatially aliassed

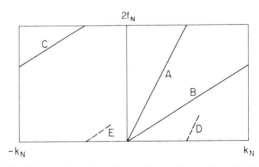

Fig. 2.54. Illustrating the effects of two-dimensional aliassing.

although if it was sufficiently temporally aliassed through lack of anti-alias filtering, it could reappear at D as shown. Event B is spatially aliassed and consequently reappears at C. If it was sufficiently spatially aliassed, it might even wrap-round temporally as event E, although this is rare in practice. So spatial aliassing is simple to identify in the $f - k$ domain. The misunderstandings seem to arise when it is required to recognise its effects in the $t - x$ domain.

In $t - x$, the symptoms of spatial aliassing are also known as dip-reversal. The act of wrapping round from $+ k_N$ to $- k_N$ corresponds to a reversal of dip. Fig. 2.55 a–c show the $t - x$ domain for a 8 Hz sinusoidal wave

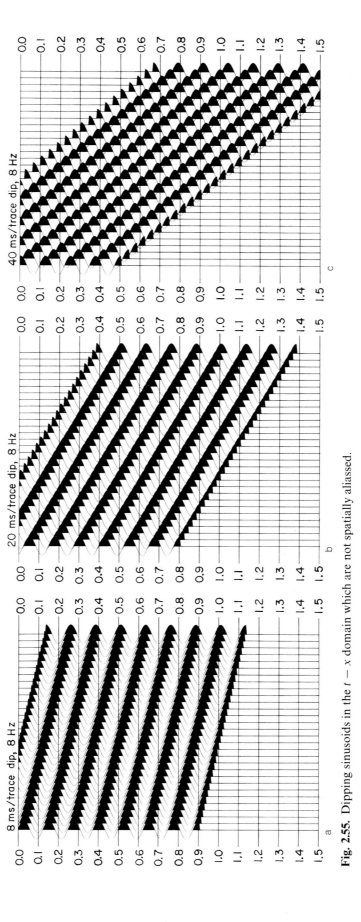

Fig. 2.55. Dipping sinusoids in the $t - x$ domain which are not spatially aliassed.

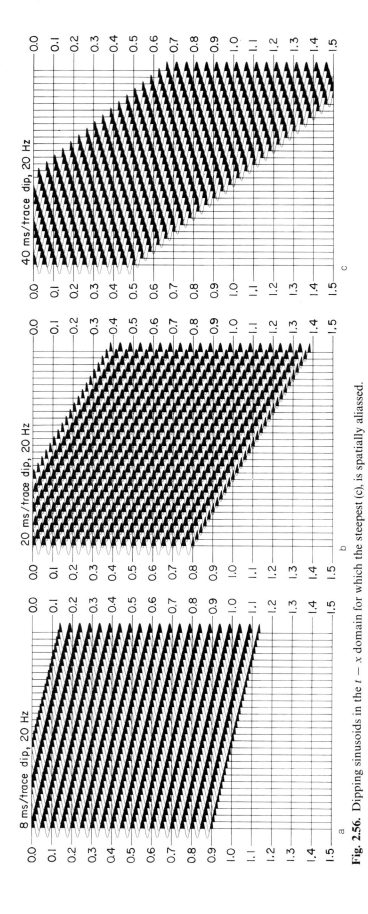

Fig. 2.56. Dipping sinusoids in the $t - x$ domain for which the steepest (c), is spatially aliassed.

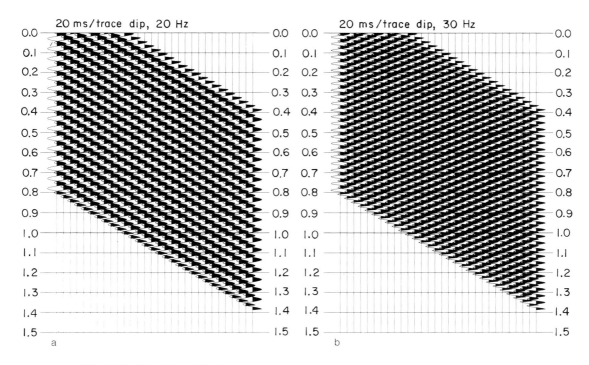

Fig. 2.57. The effects of $f - k$ filtering of two dipping sinusoids, one of which is spatially aliassed and is unaffected by the filtering. Diagram (a) shows the unfiltered form and diagram (b) shows the filtered form for this dip which also appears unfiltered.

with phase velocity such that the event dips at 8, 20 and 40 msec/trace. None of these events are spatially aliassed as can easily be seen by plotting their $f - k$ responses. The $t - x$ domain responses appear as 3 progressively steeper dips. On the other hand, in Fig. 2.56a–c, the dips in (a) and (b) are not aliassed, but the dip in (c) is spatially aliassed. Notice how it appears as a conflicting dip. The danger of course in $f - k$ filtering of aliassed data in practice, is that the dips will have changed and will no longer be filtered. Fig. 2.57a–b

shows two sinusoids of the same dip but of different frequencies, 20 and 30 Hz respectively. The 20 Hz sinusoid is not spatially aliassed whereas the 30 Hz one is. If this dip is filtered out using a $f - k$ filter, (b) is unaffected as shown. In the $f - k$ domain, Fig. 2.58 shows what has been done. The shaded zone has been filtered at A leaving the aliassed remnant at B. Of course, the filtering zone could also be wrapped round, but this would attenuate high frequencies at lower dips as in region C.

All of the relevant quantities are summarised in Table 2.10.1 for convenience.

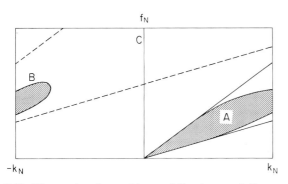

Fig. 2.58. Illustrating the problems of filtering spatially aliassed data. In this case, the continuation of the filtering fan is not applied as it will also attenuate the high frequencies in the desired data as well as the unwanted aliassed part of the noise indicated by B, the continuation of A.

Table 2.10.1

Quantity	Symbol	Dimension	Relation to other quantities
Frequency	f	time^{-1}	$f = 1/T$
Wavenumber	k	length^{-1}	$k = 1/\lambda$
Phase velocity	v_p	length/time	$v_p = f/k = f\lambda = \lambda/T$
Dip	θ	time/length	$\theta = g/v_p = gk/f$
Period	T	time	$T = 1/f$
Wavelength (spatial period)	λ	length	$\lambda = 1/k$
Geophone spacing	g	length	$g = v_p\theta = f\theta/k$

Chapter 3　　　　　　　　Seismic Data Processing

3.1 Introduction

In the very early days of exploration seismology, the geophysicist in the field did everything; acquisition, maintainance of the field equipment, making the tea, 'processing' of the data and finally interpretation. As the industry came of age, each of these activities became the job of the specialist, with the possible exception of making the tea. Seismic data processing itself experienced an enormous boost following the widespread application of the digital computer during the early to mid 1960s. From its early beginnings, the seismic exploration industry has grown into one of the largest users of scientific computing technology. Indeed, its extraordinary demands have actually led to the development and general acceptance of new devices in computer technology, for example the array processor, as described in Chapter 1 of this book. It is the second largest user of magnetic tape in the world behind the US government, and it is not uncommon to find seismic data processing installations with active, i.e. regularly used, tape libraries of 100 000 tapes and more.

Seismic data processing is an unusual concoction of highly esoteric and objective mathematical techniques in signal processing, blended with the subjective approach of the human interpreter. Indeed the whole object of seismic data processing is to massage seismic data recorded in the field into a coherent cross-section of significant geological horizons in the earth's subsurface. These data are usually severely contaminated with various kinds of noise. It is this reliance on the subjective abilities of the interpreter which gives the subject its character, and is at once its greatest strength and its greatest weakness. It is a strength because of the prodigious ability of the eye/brain to make sense of poor data which we are unable to explain. It is a weakness because this ability is largely unfettered by observational evidence and many of our comparisons and models have strong subjective content which it is impossible to verify. Interpreters thus work in a vacuum. A side-effect is that we are often unable to tell if new processing techniques are indeed valuable or have marketing content only. Many experts in industry will still agree that one of the oldest and most widely studied processes still in use, that of predictive deconvolution, does 'something nice to data, but we don't know what'. This is despite the fact that the underlying assumptions of this process are rarely, if ever, satisfied; however, more of this later.

This chapter will attempt to explain how compressional mode reflection seismic data is processed in prac-

tice, and what is going through the processor's mind at various stages. As such, it is almost entirely devoid of theory and is exemplified by copious diagrams.

The next section, 3.2, is devoted to a necessary but brief discussion of tape formats. Tape formats are agony in any language and one day there may come a time when they have historic value only.

The following three sections, 3.3–3.5, include a discussion of data processing with emphasis on the marine environment.

Section 3.6 discusses the impact of irregular and extended geometry as compared to the simplicity of 2-D marine data. This extends the relevance of the chapter to both 2-D land and 3-D land and marine processing environments.

Finally, Section 3.7 describes some important differences between land and marine processing. If this chapter is to have any theme, it is 'Happiness is a flat spectrum and a regular geometry'.

3.2 Demultiplexing, gain recovery and tape formats

3.2.1 Introduction

The subject of demultiplexing and gain recovery is eminently tedious but necessary. The fact that the subject exists at all is yet one more manifestation of the point made in Chapter 1, that the demands of seismic data acquisition and processing has stretched the electronics of the time to, and occasionally beyond, its limits. In view of the fact that the electronics industry seems to have caught up temporarily on the acquisition side, expertise in this area would seem to be a redundant skill and, with luck, the need for demultiplexing and gain recovery will disappear over the next couple of years. Consequently, a rather sketchy introduction to this topic will be given here in the fond hope that the reader will never need any of this information!

Before proceeding, it is important to point out that the gain recovery applied to seismic field tapes has nothing to do with the gains used in data processing as discussed in Section 3.3.3.

Currently seismic field tapes are recorded in a number of standard formats, SEG A, SEG B, SEG C and SEG D, corresponding to the A, B, C and D formats of the Society of Exploration Geophysicists (SEG), although a small number of contractors still stick to their own, often incomprehensible, formats. The SEG formats are documented in detail in an SEG monograph entitled *Digital tape standards*. It should be noted that S E G D is the most modern and caters for both multiplexed formats like S E G A, S E G B and S E G C and demultiplexed formats like S E G Y. S E G A is now almost extinct and S E G C is a full floating point multiplexed format which is not used often because of the extra tape required (approximately 60 per cent more than S E G B). In some circumstances, for example in land acquisition, its greater dynamic range (32 bits) is essential. As the philosophy is the same and the volume of S E G B data currently dwarfs the other alternatives, only this multiplexed data format will be considered in detail here, although it should be noted that the volume of S E G D data will increase proportionally to the increase in recording of 240 channels and more.

Demultiplexed data will be exemplified using the S E G Y format.

3.2.2 S E G B

Multiplexed data is also known as time-sequential format and demultiplexed data as trace-sequential. These two descriptions are explained in Fig. 3.1. Mathematically, the operation corresponds to a simple matrix transpose. What makes it so fiddly is the various numeric format conversions which have to be done.

S E G B format is an instantaneous floating point multiplexed format which has acceptable dynamic range in most cases, occupying effectively 20 bits or two and a half bytes per sample. The overall format is shown in Fig. 3.2. Each shot produces two records, a header record and a data record, each separated by an I R G, as defined in Chapter 1. (Individual shots are separated by an E O F.)

3.2.2.1 The header record

A typical S E G B header is shown as a hexadecimal (base 16, with A–F corresponding to 10–15) dump in Fig. 3.3 and consists of three parts:

General header constants These include a variety of things in the first 24 bytes. The information is decimal encoded in binary form and is packed as two characters per byte, as will be illustrated shortly. A number of

these constants are distinctly esoteric and only the following are of significance, their values from Fig. 3.3 being included in parentheses.

Characters

1–4	File number	(363)
21–23	Bytes per scan	(314)
24	Sample interval in msec	(4)
33–34	Record length	(6)
36	Type of record	(8 = normal shot)
37–38	Low cut filter frequency	(8)
47	Alias filter frequency 3 bd point	(4 = 62.5 Hz)

Gain header constants These contain the fixed and early gain constants for all the data and auxiliary channels. These are simple multipliers which should be applied to all the samples for any particular channel. In Fig. 3.3, there are 120 data channels with a fixed channel as a data channel. Note however, that the early gain code is applicable only to S E G A and in this case is therefore 00.

Extended header constants There is no fixed definition for these cells but they usually contain such information as cable compass values for 3-D surveying and a variety of other things. In Fig. 3.3, everything after byte 278 (24 bytes of header, 250 bytes of fixed and early gain codes for 120 data channels and 5 auxiliary channels, and four bytes to finish the main header) is extended header information.

3.2.2.2 The data record

An example of part of a data record corresponding to the header of Fig. 3.3 is shown as a hexadecimal dump in Fig. 3.4. Individual scans of each channel at every time sample consist of 314 bytes as detailed by the header record, including a 4 byte sync code. This is a recognisable string of 4 bytes or 8 characters which the demultiplex program can recognise and thereby determine the start of the next scan. In order not to confuse the demultiplexing operation, this code must not be a

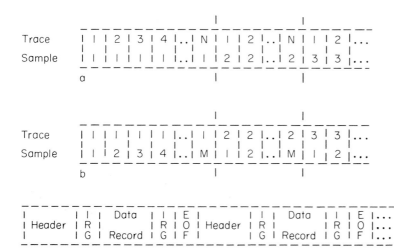

Fig. 3.1. Illustrating the different organisation of (a) multiplexed tape formats, and (b) demultiplexed tape formats.

Fig. 3.2. A schematic of the S E G B multiplexed tape format.

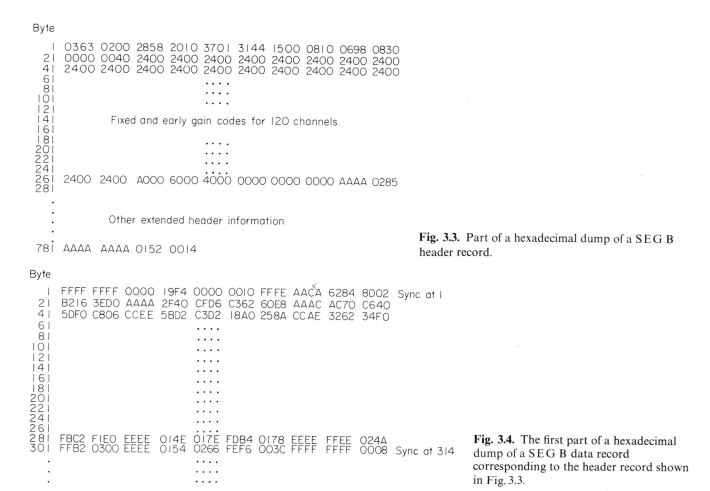

Fig. 3.3. Part of a hexadecimal dump of a SEG B header record.

Fig. 3.4. The first part of a hexadecimal dump of a SEG B data record corresponding to the header record shown in Fig. 3.3.

possible data value. Since SEG B has an effective 15-bit mantissa, the least significant (rightmost) bit is always zero. The sync code is therefore designed to have the least significant bit of each of its four bytes set to 1. Apart from this restriction, the other bits may be set to anything. On the data of Fig. 3.4, the sync scan is FFFF FFFF.

To finish the discussion of SEG B, a number from the data record will be converted into its equivalent number of volts, thus illustrating the process of gain recovery, whereby the SEG B format numbers are converted to full floating point format, often SEG Y (IBM floating point) format. Noting that the first 10 bytes after the sync code contain the 5 auxiliary channel sample values with no embedded instantaneous gain information, inspection of Fig. 3.4 reveals the following string of characters:

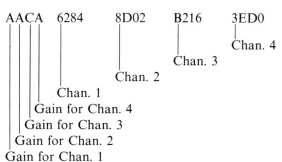

As can be seen, the fact that the rightmost bit is zero in SEG B makes all the channel data values appear even.

Channel 1 will be used as an example. To calculate the actual sample value for channel 1, sample 1, simply do the following:

An SEG B sample value is defined to be

$$2^F \times C \times 2^{-I}$$

where F is the fixed gain

C is the channel value expressed as a binary decimal with the binary point to the left of the most significant (leftmost) bit, excluding the sign bit in the first position

I is the instantaneous gain.

In this case, $F = 4$, $C = .6284_2$ and $I = 10$ (hexadecimal A). Here the suffix 2 denotes base 2.

Now, $.6284_2 = 6284_2 \times 2^{-15}$

$$= 25220_{10} \times 2^{-15}$$

Hence, the sample value is

$$2^4 \times 25220_{10} \times 2^{-15} \times 2^{-10}$$

$$= .01217$$

and so on . . . .

It would be nice to think that this figure is simply volts. This, unfortunately, is not always the case as it depends on the analogue to digital converter and sometimes involves a factor of two, for example the units might be half-volts.

3.2.3 SEG Y

This format is quite nice if the computer on which the tape is being read is an IBM machine as both the character and numeric formats are based on this manufacturer's machines. Unfortunately, the IBM character set, EBCDIC (Extended Binary Coded Decimal Interchange Code) is not an international standard, the only such character set being ASCII (American Standard Code for Information Interchange) which almost every other computer system uses. Consequently, SEG Y is an inefficient and unnecessarily fiddly format in general.

In addition, it is somewhat aged and not designed for modern requirements like 3-D and so on; however it does provide a convenient, simple method for interchanging datasets as virtually all computer systems in the seismic industry have software capable of reading this format.

The format is indicated pictorially in Fig. 3.5.

```
 ___ ___ ___ ___ ___ ___ ___ ___ ___
I  Tape reel  I I I  Data   I I I  Data   I I I...
I   Header    I R I Record I R I Record I R I...
I   Record    I G I    I   I G I    2   I G I...
 ___ ___ ___ ___ ___ ___ ___ ___ ___
```

Fig. 3.5. A schematic of the SEG Y demultiplexed format.

3.2.3.1 The tape reel header record

The tape reel header is divided into two parts:

1 A 3200-byte EBCDIC skeleton arranged as 40 card images which at least one of the authors (LH) has never seen filled in correctly!
2 A 400-byte binary coded block, of which 60 are assigned and 340 left available for future expansion. Unfortunately, this format was designed in the days when optimal usage of computer memory space was desirable and so many of the 4-byte fields contain more than one piece of information packed together to fit.

Of these 60 bytes, the following are recommended as being the minimum set to be correctly specified:

Bytes
- 5–8   Line number
- 9–12   Reel number
- 13–14 Number of data traces per record
- 15–16 Number of auxiliary traces per record
- 17–18 Sample interval in microseconds for this reel
- 21–22 Number of samples per data trace for this reel
- 25–26 Data sample format code. Almost invariably 1 for full floating point.
- 27–28 CMP fold
- 55–56 Measurement units. 1 = metres, 2 = feet.

The tape reel header is separated by an IRG from the first data record.

3.2.3.2 The data record

The first 240 bytes of each data record are binary encoded trace identification information of which the following are the recommended minimum to be set correctly:

Bytes
- 1–4   Trace sequence number within line
- 9–12   Original field record number
- 13–16   Trace number within original field record
- 29–30   Trace identification code
  - 1 = seismic data
  - 2 = dead
  - 3 = dummy
  - etc.
- 115–116 Number of samples in this trace
- 117–118 Sample interval in microseconds for this trace.

The remaining information in the record pertains to the sample values, almost invariably encoded in IBM floating point format. This format employs a 7-bit exponent and 24-bit mantissa. The exponent is in excess 64 format.

If the bit value in position i is denoted as b_1, where $i = 1$ corresponds to the most significant (leftmost) bit and $i = 32$ to the least significant bit, the data value is represented by the following floating point number:

$$16^{(b_{2-8} - 64)} \times \sum_{j=1}^{6} \frac{b_{j, j+1, j+2, j+3}}{16^j}$$

where b_{2-8} denotes bit positions 2–8 taken as a binary number. This number is negative if b_1 is 1 and positive otherwise.

This rather formidable looking object can be exemplified by considering the hexadecimal 32-bit number 41100000. In this case,

$$b_{2-8} = 41_{16} = 65_{10}$$

making

$$16^{(b_{2-8} - 64)} = 16^{(65 - 64)} = 16$$

and the horrible looking summation is

$$\left(\frac{1}{16^1} + \frac{0}{16^2} + \frac{0}{16^3} + \frac{0}{16^4} + \frac{0}{16^5} + \frac{0}{16^6} \right)$$

$$= \tfrac{1}{16}$$

Hence, the whole thing boils down to

$$16 \times \tfrac{1}{16} = 1$$

3.3 Fundamentals of marine seismic data processing

3.3.1 The mid-point assumption

In the seismic reflection method, sound waves are generated which then travel down into the earth. In general, for marine acquisition, the source of acoustic energy is made as impulsive as possible and usually derives from the release of compressed air (air-gun), rapid acceleration of water (water-gun) or electrical discharge (sparker). On land, the source is usually vibrational (Vibroseis*) or explosive (dynamite). A proportion of the generated energy reflects from interfaces between different lithologic units according to the contrast in acoustic impedance as defined in Chapter 2. The returning echoes are received by hydrophones (marine acquisition) or geophones (land acquisition) and the resulting recordings digitised and placed on tape. The reflecting points in the mid-point assumption are deemed to lie half-way between the source point and the receiver point using Snell's first law. The geometry is shown in Fig. 3.6. This assumes that the earth consists of plane homogeneous layers.

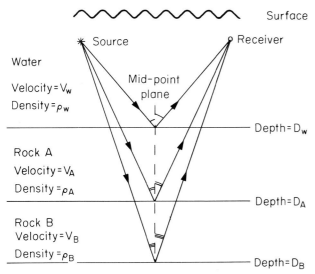

Fig. 3.6. Mid-point ray-paths for a horizontally layered medium, ignoring the effects of Snell's second law.

The strength of the echo for normal incidence from a particular interface is governed by the reflection coefficient which is the ratio of the amplitude of the displacement of a reflected wave to that of the incident wave, i.e.

$$R_{12} = \frac{a^r}{a_i}$$

The subscripts 1 and 2 refer to the adjacent layers separated by the interface. R_{12}^2 is therefore the ratio of reflected-to-incident energy. The transmission coefficient can be defined as

$$T_{12} = \frac{a_i + a_r}{a_i} = 1 + R_{12}$$

Fig. 3.7 exemplifies.

An alternative formulation using simple wave theory, shows that the reflection coefficient is related to the acoustic impedances of each side of the interface as follows:

$$R_{12} = \frac{\rho_2 V_2 - \rho_1 V_1}{\rho_2 V_2 + \rho_1 V_1}$$

where ρ is density and V is compressional velocity, with an equivalent expression for T_{12}.

* Trade mark of Conoco.

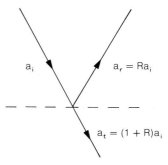

Fig. 3.7. Reflection and transmission coefficients at an interface.

In the simple model used in Fig. 3.6, Snell's second law was ignored. Incorporating this, the extended model is shown in Fig. 3.8. The ray paths are shown refracting away from the vertical corresponding to the normal trend for increasing acoustic velocity with layer depth. An interesting insight into Snell's second law is obtained by considering it as an expression of Fermat's principle which states that wave energy will always travel along a least time path. Hence, although the ray paths reflected from depth D_A and D_B are longer than their equivalent in Fig. 3.6 due to the inclusion of refraction, a correspondingly greater proportion of their length lies in the faster velocity layers resulting in a minimum travel time.

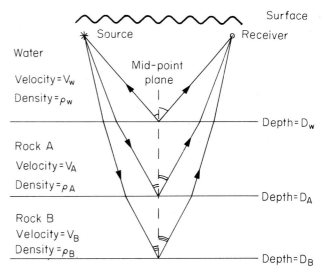

Fig. 3.8. As Fig. 3.6 but including the effects of Snell's second law.

An unfortunate aspect of this latest model is that it assumes that lithologic units in the earth are flat, whereas the reality is very different. A more realistic model still must therefore include the possibility of dipping layers as shown in Fig. 3.9. It can be seen from this latest refinement, that the reflecting points no longer lie in the mid-point plane. Whilst this calls into question the validity of the mid-point assumption, the offset between source and receiver is usually small compared with the depth of reflecting boundaries of interest. The offset of reflecting points from the mid-point plane is small therefore, and at least in areas where dips are

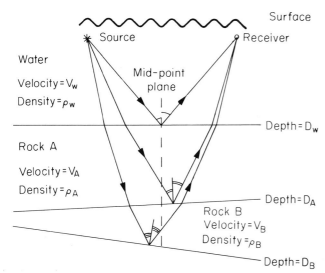

Fig. 3.9. As Fig. 3.8 but including the effects of a dipping layer.

gentle, travel times are close to those expected for a horizontally layered earth. Although the inadequacies of the assumption should always be borne in mind in areas of complicated geology, the concept of a mid-point plane allows the generation of multi-coverage data with the resulting benefits of improved data quality, attenuation of noise and the ability to estimate subsurface velocities.

3.3.2 Multi-coverage common mid-point (CMP) data

A survey of the subsurface by movement of a single source and receiver can be made such that a recording is collected, say, every 25 metres. This type of coverage is referred to as single fold or one hundred per cent coverage and would be collected by means of a boat towing a single source and receiver as shown in Fig. 3.10. A shot fired every 25 m would result in reflected energy being recorded for mid-points 25 m apart.

4000 metres long. Fig. 3.11 shows a schematic of a 6 channel cable.

Geometrical consideration of a single shot into a multi-channel cable shows that the spacing of mid-points cannot be more than half of the receiver group spacing. Dependent on the distance (pull-up) between shots, recorded traces share common mid-points which are separated by, most commonly, half the surface group interval or occasionally a simple fraction of it.

It is often instructive to construct a chart which shows the mid-points of all source–receiver pairings within each shot on a line. This display, usually known as a (subsurface) stacking chart, is shown in Fig. 3.12 for the case of a six trace cable with 25 m receiver group spacing and 12.5 m shot spacing. It can be seen that for this acquisition configuration, each recording channel supplies one data trace from consecutive shots to a CMP gather, and 600 per cent coverage results. This applies to all CMP gathers except:

1 The very beginning and ending of each line where fold of coverage is gradually built up and drops off. These zones are known as the roll-on or taper-on, and roll-off or taper-off respectively.

2 When a shot is missed, or rejected after acquisition on grounds of poor quality. Fold of coverage in this case will drop by one over a 6 CMP zone.

3 It is common in reconnaissance processing to perform some kind of data compression very early on in the processing sequence for reasons of economy. Two kinds are in popular usage:

Vertical or file sum: equal offset traces are summed from adjacent shots.

Horizontal or trace sum: adjacent traces within the same shot are summed.

In each case, the total number of traces is halved. Fold of coverage changes dependent on the summation technique used.

The stacking chart allows the geophysicist to define exactly which traces from different shots have common

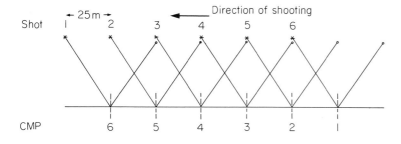

Fig. 3.10. A schematic of 100 per cent coverage or single-fold data.

An improvement in data quality (and economy!) by means of averaging can be achieved when many receivers are towed behind the boat on a cable, and care is taken that the spacing of shot and receiver locations coincides in a regular manner. In practice, a group of hydrophones with considerable spatial extent (up to 100 metres or so) are fed into a single recording channel and distances measured to the middle of this group. Any number of channels from twelve to a thousand are presently in common use on cables ranging from 300 to

mid-points, and which therefore can be averaged or stacked. Most processing systems allow the option to automatically generate a stacking chart either prior to, or during, the reorganisation of data traces from common shot gathers to common mid-point gathers. As might be expected, every trace in a CMP gather will originate from a different offset receiver. This fact gives the CMP method of acquisition an additional important feature: by means of mathematical comparison of traces within the CMP gather, an acoustic velocity

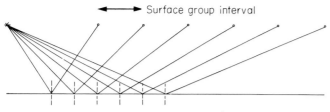

Fig. 3.11. A schematic of a 6 channel cable.

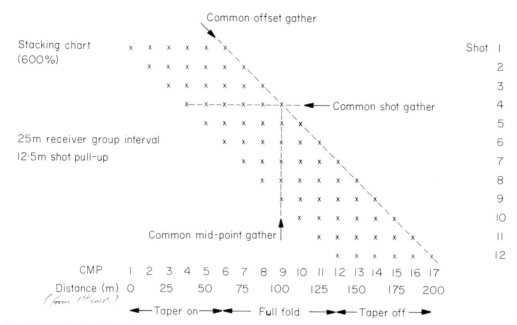

Fig. 3.12. A subsurface stacking chart for a 6 trace cable with a 25 m receiver group interval and a 12.5 m shot spacing.

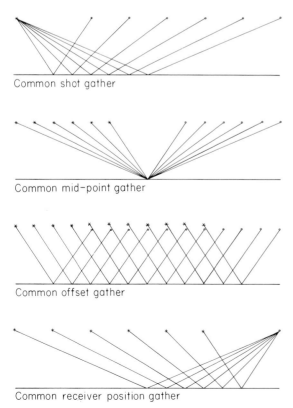

Fig. 3.13. Common collectives for seismic traces.

profile can be estimated. Certain modern data processing techniques may operate in any one of the domains which are showed in both Figs. 3.12 and 3.13.

Efficient practical processing can only be achieved if the geophysicist possesses a solid understanding of the consequences of different acquisition geometries. (The exercises on p. 166 are intended to help the reader in this respect.)

3.3.3 The nature of seismic traces

3.3.3.1 The convolutional model revisited

So far, only the geometry of acquisition has been considered without investigating the nature of the signal received by a hydrophone or geophone. Here, the discussion which led to the convolutional model in Chapter 2 will be expanded pictorially to fill in the geophysical details.

Fig. 3.14 shows the reflection strength for some portion of an idealised recording plotted against two-way travel time, which is the time taken for a seismic pulse to travel from the source to the reflecting interface and back up to the receiver. The recording itself is known as a seismogram or seismic trace, a terminology which dates back to analogue recording days

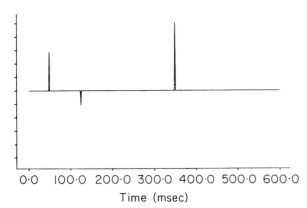

Fig. 3.14. A synthetic reflection series.

when recordings were made as a trace of a pen on a rotating drum. This particular trace represents three interfaces of variable impedance contrast, one of which results in a negative reflection coefficient. A negative reflection coefficient occurs at a decrease in acoustic impedance across an interface. Since density is generally quite slowly varying in regions in which exploration seismologists are interested, this usually implies a decrease in acoustic velocity. The timing of reflections from a specific interface will vary from trace to trace according to average velocity down to the interface, depth of the interface, and offset of receiver from source. This reflection series determines the timing and relative strengths of the reflected seismic pulses according to the convolutional model discussed in Chapter 2. A typical simplified seismic pulse is shown in Fig. 3.15.

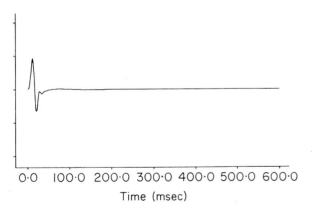

Fig. 3.15. An idealised seismic pulse.

Unfortunately, in practice, the seismic trace also contains contributions of an undesirable nature, namely:

    (a) Multiple reflections
    (b) Coherent noise
    (c) Random noise

These pollutants will now be discussed in turn.

(a)   Multiple reflections

In addition to energy returning via a simple reflection, more complex ray paths are possible as shown in Fig. 3.16. As reflection from more than one interface is involved, these returns are collectively known as multi-

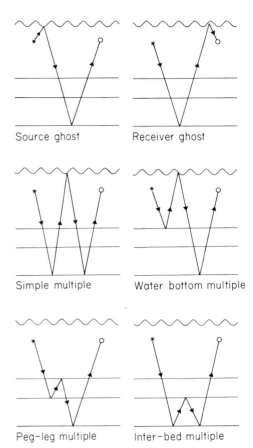

Fig. 3.16. Ray paths of some common multiple families.

ples. For all but very shallow interfaces, it can be reasonably assumed that travel paths are vertical. The additional distance travelled along these multiple ray paths then results in a constant time delay independent of receiver offset.

The upper two examples in Fig. 3.16 are generally considered separately from other multiple categories due to the short time delay involved and high reflection coefficient of the surface boundary which approaches −1. Such ghost reflections were considered briefly in Chapter 2. Ghost reflections produce a high amplitude pulse of negative polarity which follows so closely behind the primary pulse as to interfere with it and produce one modified pulse. In practice, of course, both source and receiver ghosts must be included. Assuming a source depth of 7.5 m and a receiver cable depth of 9 m, typical values in the marine environment, the modified source pulse can be calculated by convolving the original with the function shown in Fig. 3.17. In this case, a reflection coefficient of − 0.9 is shown taking into account various factors such as surface roughness and anelasticity of the surface for high-powered sources. The resulting modified source pulse is shown in Fig. 3.18 and as can be seen, now bears little resemblance to the original of Fig. 3.15. Convolving this modified pulse with the original reflection series of Fig. 3.14 results in the primaries-only trace of Fig. 3.19.

The other complex ray paths illustrated in Fig. 3.16 generate additional reflections of the modified pulse at time lags dependent on interface depth and separation. These reflections can occur at time lags which vary from very short, in the case of a shallow water-bottom multiple, to very long in the case of simple multiples of deep interfaces. Their existence often creates serious problems in data processing, and techniques for their attenuation will be discussed later. Briefly, however, these are based on two properties of a multiple:

1 The average velocity of the pulse is usually less than that of a simple reflected pulse arriving at the receiver at the same time (velocity discrimination).

2 The time lag of the multiple reflection is predictable, usually from analysis of the data, allowing it to be attenuated using deconvolution techniques.

Again, multiples can be incorporated into the seismic trace by convolution with a simple time series. A time series which will generate two water-bottom multiples is shown in Fig. 3.20. In this case, a water depth of 150 m and water bottom reflection coefficient of 0.5 were used. The resulting primaries plus water-bottom multiples trace is shown in Fig. 3.21.

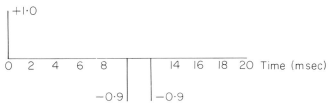

Fig. 3.17. A primary plus source–receiver ghost triplet.

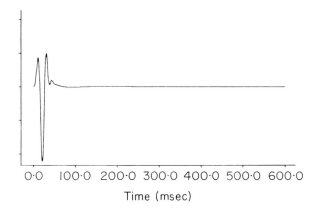

Fig. 3.18. The modified source pulse including source and receiver ghosts. This pulse is the convolution of the responses of Figs. 3.15 and 3.17.

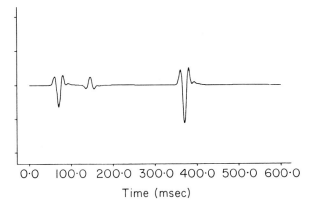

Fig. 3.19. A band-limited primaries-only synthetic seismogram formed by convolving the pulse of Fig. 3.18 with the reflection series of Fig. 3.14.

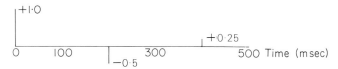

Fig. 3.20. An idealised water-bottom multiple reflection series.

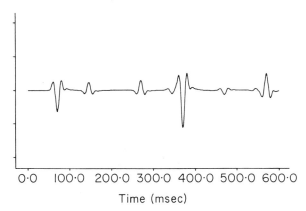

Fig. 3.21. A primaries plus water-bottom multiples synthetic seismogram.

(b)   Coherent noise

The following are usually included in the category of coherent noise:

Direct waves: energy travelling directly through the water to the receiver array.

Refracted waves: energy striking an interface at the critical angle, followed by refraction along the interface before scattering back into the receiver array.

Diffracted waves: energy scattered in all directions from a point scatterer, not necessarily located in the plane of the survey. Seabed irregularities, for example, sunken wrecks, are a major source of such energy.

Vibrational noise: energy resulting from tug of the boat and the tail buoy on the cable in roughish seas, as well as boat noise including source generated noise.

Interference: energy generated by other vessels, especially seismic survey vessels which can be a serious problem even a 100 kilometres away.

The common characteristic of this kind of noise is the line-up of energy in a distinguishable manner across the traces of a shot gather as shown, for example, by Larner *et al.* (1983). Exploitation of this with two-dimensional filtering techniques provides the possibility of attenuation. The pulse shape of coherent energy as recorded by the horizontally extended receiver array is, in general, dissimilar to the reflected pulse shape, even if source generated, due to the directionality of the receiver array. The appearance of coherent energy on a single trace therefore is unpredictable and not necessarily localised into a recognisable pulse shape.

(c)   Random noise

By definition, random noise exhibits no correlation from trace to trace, and no specific noise amplitude can be predicted by knowledge of the generating mechanism. Often, however, information about its statistical behaviour can be calculated. Mathematical analysis of the data can result in RMS amplitude and frequency spectral estimates. These parameters may vary with shot, receiver group and two-way travel time.

Some of the more common noise generating mechanisms are:

Instrument noise: thermal motion of electrons in recording equipment electronics; interference.

Machinery: compressors and other heavy machinery on board the seismic vessel.

Power lines: generally 50 or 60 Hz, can be picked up from the seismic vessel or buried lines.

Cable noise: motion of the cable through the water, wave action, leakage.

Improvements in the design of seismic vessels, instruments and cables are continually reducing the effect of the mechanisms mentioned above. This has reached the point where the advent of very powerful sources and the resulting source generated noise usually leads to this dominating background noise levels. Signal to noise improvement obtained at depth by the use of large arrays of sources has often been disappointing. This can be attributed to the comparable increase in source generated noise as is shown by the work of Larner *et al.* (1983), and Newman (1984). Methods of random noise reduction are based on averaging techniques and frequency filtering and will be discussed later.

The simple convolutional trace with multiples and added random noise will look something like Fig. 3.22.

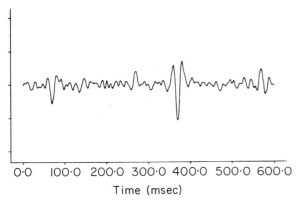

0·0   100·0   200·0   300·0   400·0   500·0   600·0
Time (msec)

Fig. 3.22. A synthetic seismogram including the effects of primaries, water-bottom multiples and additive noise.

3.3.3.2 Amplitude decay

Seismic traces as displayed in their demultiplexed and gain-recovered raw state show a very marked decrease in amplitude with time. Such data are known as 'geophone amplitude' data and contain the various attenuation effects at work on the seismic pulse as it passes through the earth.

The range of amplitudes recorded is of the order of 1 to 10 000 or 100 000 which corresponds to 80 and 100 db respectively, as defined in Chapter 2. Recording equipment is therefore designed and periodically tested to produce instrument noise of more than 80 or 100 db down on maximum setting.

Loss mechanisms are varied but divergence (geometrical spreading according to the inverse square

law), absorption by anelastic propagation, scattering from small heterogeneities and transmission losses are generally accepted as being the major contributing factors.

(a)   Divergence loss

Consider a constant velocity medium for which ray paths are therefore straight, and in which energy passing through unit area at distance r from the source is proportional to r^{-2}. The recorded pressure amplitude is therefore proportional to r^{-1} and compensation for 'spherical' spreading can be made using a scaling factor proportional to r, i.e.

$$G(t) \propto r \propto Vt$$

When a layered medium of variable velocity is considered and refraction allowed, as Newman (1973) showed, the velocity required is the RMS velocity. A further factor of V also results, giving

$$G(t) \propto V_{RMS}^2(t)t$$

The proportionality constant is arbitrary and can be selected from some reference level, for example,

$$G(t) = \frac{V_{RMS}^2(t)t}{V_{RMS}^2(t_0)t_0} \qquad (3.3.1)$$

Figure 3.23 illustrates.

(b)   Transmission loss

Energy transmitted through the earth is reduced by partial reflection at interfaces according to impedance contrasts, as discussed earlier. The severity of this loss mechanism is dependent on geology and is difficult to model even when well logs are available. Cyclic sequences of contrasting impedance can cause major losses, whereas thick layers of constant character providing a gradual velocity increase with depth will have a lesser effect.

Transmission loss can be overestimated when substantial thin layering is present. Short delay interbed multiples will restore lost amplitude to a certain extent, but distort the pulse shape.

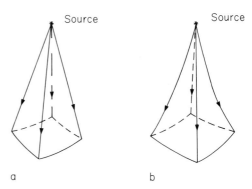

Fig. 3.23. Illustrates geometric spreading in a constant velocity medium (a) and a layered medium of variable velocity (b).

(c) Absorption

As the seismic pulse travels through the earth, energy is converted into heat through anelastic movement of rock particles. Again this mechanism is a function of geology. Very rough grains will absorb more energy than finely concentrated grains.

Absorption is heavily dependent on frequency. As we know from the behaviour of sound in air (e.g. thunder), high frequencies are attenuated more than low frequencies. A common and reasonable assumption is that loss will be a constant proportion for each cycle of a wave independent of the frequency. This implies that absorption is functionally dependent on the exponential of the number of cycles in a period of time. This along with a proportionality constant borrowed from electrical engineering known as Q (for quality, the absorption factor), is defined such that:

$$\frac{A(t)}{A(0)} = \exp\left(-\frac{\pi f t}{Q}\right) \qquad (3.3.2)$$

Physically, Q is the ratio of energy stored to energy lost per cycle. After Q cycles, amplitude falls to approximately 4 per cent of its original value. Common values of Q vary from 50 to 300, highly absorptive rocks having lower Q values. Water has a very high Q and transmits sound very well as has already been alluded to.

(d) Scattering

Irregularities and inhomogeneities scatter energy in unpredictable directions reducing the amplitude of the transmitted pulse. This effect is geology dependent and becomes important in highly fractured zones.

(e) Mode conversion

When a compressional wave is incident at an interface at other than normal incidence, mode conversion to shear waves and other more complex modes may take place. Significant conversion takes place at high incidence angles when a large horizontal component of particle movement is available to generate reflected and refracted shear waves. Shear waves, of course, cannot travel through water, although energy may arrive at the hydrophones which has spent some part of its time as shear wave energy before being reconverted.

The mechanisms outlined above all contribute to a decay with time in the amplitude of reflected signal received at the cable. Some noise contributions (e.g. cable generated noise) do not vary with time whilst others (e.g. diffraction from near surface anomalies) decay very slowly. The result is that after a certain two-way time, traces become dominated by noise which shows little further decay as is shown by Fig. 3.24, an analysis of amplitude decay averaged over the near offset traces of some shallow high resolution data.

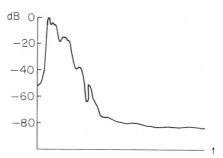

Fig. 3.24. The amplitude decay of an average of the near offset traces of Fig. 3.29 as a function of two-way travel time.

3.3.3.3 Decay compensation

Many processing techniques rely heavily on the extraction of statistical information from stationary time series. Roughly similar amplitude levels throughout the data are therefore required. It is also important for quality control purposes to be able to bring all the data within a displayable dynamic range. Equalisation techniques vary considerably but fall into two categories.

(a) Data independent equalisation

Very few, if any, equalisation procedures are truly data independent but this heading is used to describe those where a large amount of data is used to derive a trace invariant scaling curve. The scaling of any particular sample will therefore be independent of its value or the average value of those samples around it.

The amplitude level of a zone is usually measured as a mean absolute or RMS amplitude. Figure 3.25 shows two absolute amplitude measures. RMS amplitude is more heavily influenced by large individual sample values than the mean absolute amplitude. This can be important when the data contains high amplitude reflections or noise bursts.

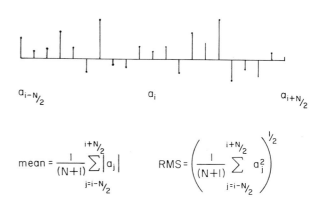

Fig. 3.25. Two common measures of average amplitude.

Amplitude analysis of a single trace can be achieved by computation of mean amplitudes over contiguous zones, commonly 20–200 msec long, and interpolation between points as shown in Fig. 3.26.

Overlapping time zones are often used to reduce the interpolation required, sometimes to the extent that a zone is constructed for each sample value. To produce a

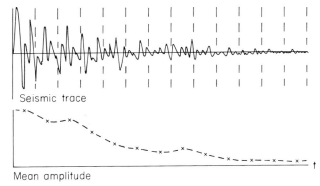

Fig. 3.26. Amplitude analysis of a single trace using the averages of a number of adjacent windows with interpolation between.

decay analysis applicable to the whole line (or survey), the individual analyses are averaged over many traces.

Compensation to a desired base level is now a simple matter of applying a scaling curve which is proportional to the inverse of the amplitude analysis curve as shown in Fig. 3.27. As an alternative to using a completely empirically determined curve, an analytical function which provides a good approximation to it can be used.

Exponential functions are a possibility with the option to hold the scalar constant when noise becomes dominant.

(b)   Data dependent equalisation

When amplitude variation is severe and unpredictable, as in the case of excessive coherent noise or complex near-surface geology, equalisation by means of a single space-invariant time function may not be sufficient. Application of the inverse curve on a trace-by-trace basis is then advisable. Methods of scaling or gaining in

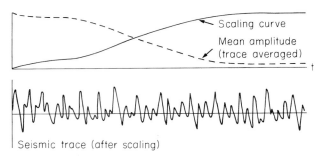

Fig. 3.27. Gain compensation.

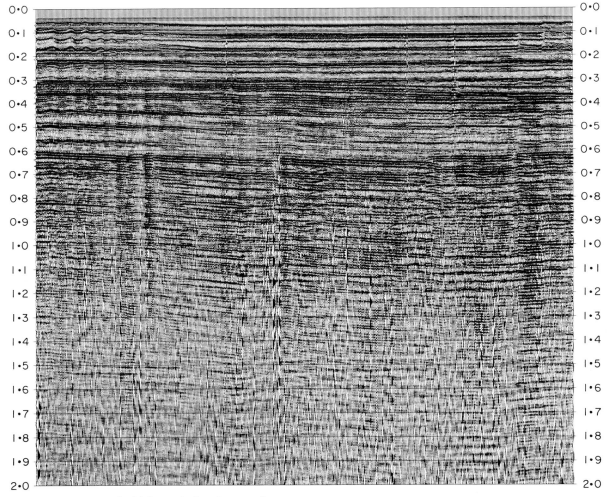

Fig. 3.28. The near traces of a high-resolution dataset after general decay compensation. This dataset will be subjected to a number of standard techniques for illustration in this chapter.

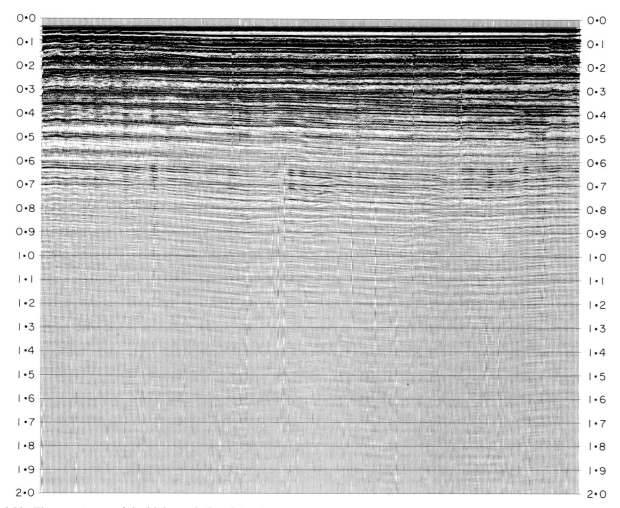

Fig. 3.29. The near traces of the high-resolution dataset without any equalisation.

which a particular sample is multiplied by a scalar derived from a window of data about the sample are categorised as automatic gain control (AGC). The important parameter which determines the severity of equalisation is usually the window size, although the flattening effect is sometimes reduced by techniques such as input feedback.

Data dependent equalisation is preferably left until a late stage in processing, as lateral amplitude relationships, which are important in stratigraphic interpretations, are destroyed, and signal to noise degraded to some degree. Figs. 3.28–3.30 show near offset traces after general decay compensation, no equalisation, and AGC with a 400 msec 'window' respectively. This particular dataset will be used to illustrate a number of standard processing techniques throughout this section.

3.3.4 Patterns within a marine shot file

A collection of traces from one shot is often referred to as a file. This should not be confused with the computer science term file which was defined in Chapter 2. Traces from shot files and seismic traces in general are invariably displayed side by side, such that two-way time increases with depth along the vertical axis. Traces are

ordered horizontally according to receiver position and average excursion amplitude of the individual traces adjusted to enable energy patterns to be detected. Even a shot file from a good seismic area, however, may appear to be an indiscernible jumble of energy line-ups as is shown by Fig. 3.36, but from a knowledge of the acquisition geometry and likely geological features, many of the energy returns can be identified. Understanding the characteristic shape of these returns as seen in a shot file is vital.

(a) Reflections

Consider the travel paths of reflected pulses when source and receiver are first co-incident and then separated by distance X. The reflecting boundary is flat and travel paths are completely within a constant velocity medium. A simple analysis of Fig. 3.31 reveals that the additional delay ΔT, or normal move-out, for a receiver with offset X compared with one at offset zero is given by

$$\Delta T = \left(T_0^2 + \frac{X^2}{V^2} \right)^{1/2} - T_0 \qquad (3.3.3)$$

This equation is of great importance in reflection seis-

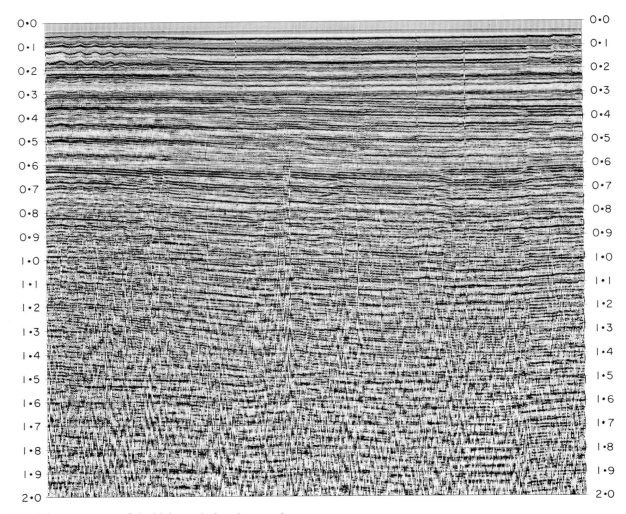

Fig. 3.30. The near traces of the high-resolution dataset after AGC with a 400 msec window.

mology and is known as the normal moveout or NMO equation. Timing corrections using ΔT, as calculated from the above, are made prior to averaging the data in the CMP stack. It is also used in the estimation of velocities.

Equation (3.3.3) represents a hyperbolic relationship between ΔT and X and should appear as such in a shot file display. The detailed shape of the hyperbola is dependent on the values of T_0 and V. It should be noted that when interfaces are at later time and/or velocity increases, this hyperbolic relation is flatter. Fig. 3.32 illustrates this point.

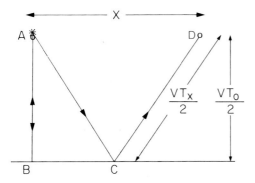

Fig. 3.31. Normal moveout (NMO).

(b)   Direct arrivals

This corresponds simply to energy which travels directly from the source to the receiver as shown in Fig. 3.33. The arrival time therefore varies linearly with receiver offset according to

$$T_x = \frac{X}{V_w} \tag{3.3.4}$$

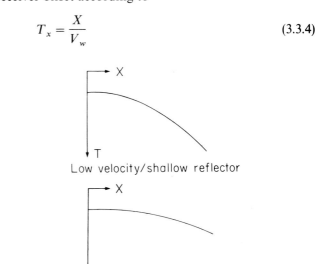

Low velocity/shallow reflector

High velocity/deep reflector

Fig. 3.32. Illustrated the flattening of the moveout curve with increasing time and velocity.

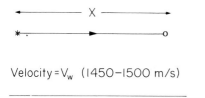

Fig. 3.33. The direct arrival. V_w is the water velocity.

The shape of the recorded pulse is usually more extended than reflected pulses as energy arrives along the axis of the receiver array.

(c) Refractions

When the energy incident on a boundary reaches the critical angle, the angle of refraction becomes 90 degrees and energy travels parallel to, and just below, the boundary as is illustrated in Fig. 3.34. This energy is refracted back through the interface and is received at all points offset greater than X_0. Delay of the pulse is linear with receiver offset and dependent on V_A only.

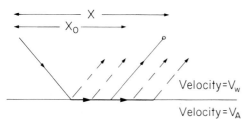

Fig. 3.34. The geometry of refraction returns.

To illustrate the above features, Fig. 3.35 is a schematic of an idealised marine shot file with examples of each of these patterns. Consider the arrivals into groups 1, 5, 12 and 22 of the 24 group cable.

Group 1 The direct arrival is usually, but not always, the first energy received at the inside group. The reflection from the water bottom (ray path $SA1$) arrives at a later time as dictated by the NMO equation.

Group 5 Energy is incident at the water bottom at the critical angle and refracted energy as well as reflected is therefore received (ray path $SB5$).

Group 12 At this offset, energy refracted along the water bottom arrives at the receiver simultaneously with the direct arrival. Although the travel path of the direct arrival ($S12$) is shorter than that of the refraction ($SBD12$), the extra distance spent in the higher velocity rock by the refracted energy results in equal arrival times. At greater offsets than this, refracted energy will be received first.

Group 22 Direct arrival energy (ray path $S22$) and reflected energy ($SE22$) arrive at the receiver group almost simultaneously. Refracted energy reaches receivers first for far offset groups.

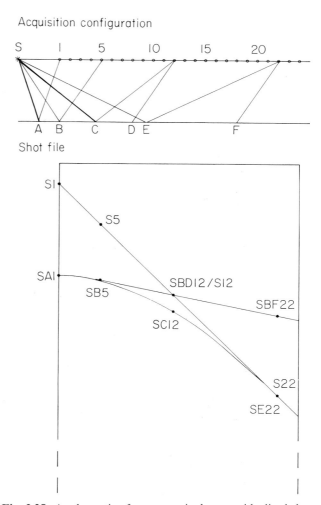

Fig. 3.35. A schematic of energy arrivals on an idealised shot file.

Figs. 3.36–3.38 show a series of shot files corresponding to the near trace sections of Figs. 3.27–3.29. Note that the first and last file shown are plain noise files, i.e. recorded without a shot being fired.

3.3.5 Common mid-point stacking

The cable geometry and distance between consecutive shots are designed such that many different source and receiver positions share the same common mid-point. There is therefore considerable redundancy whereby the estimate of the reflected signal can be improved.

3.3.5.1 The conventional mean amplitude stack

Considering a single trace from the common mid-point gather, each sample value can be regarded as an estimate of the primary reflected signal at a particular two-way travel time, contaminated by random noise. The redundancy of information available can be used to reduce the error in this estimate by taking a statistical average. The estimator most commonly used is the mean.

After correction of the data for normal moveout differences, the mean sample value at each two-way travel time can be calculated. Note that correction is invari-

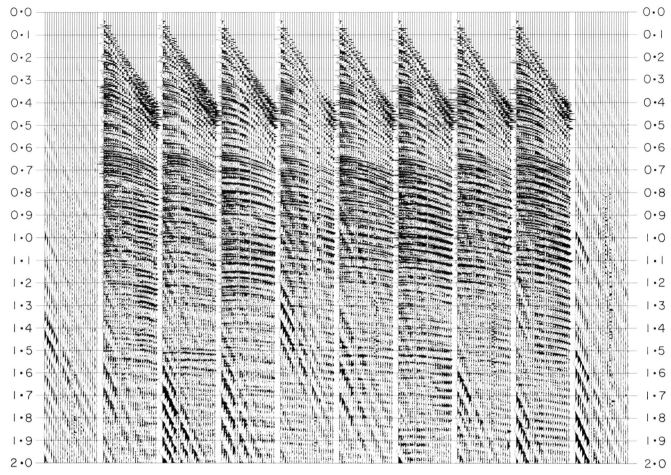

Fig. 3.36. Selected shot files from the high-resolution dataset with general decay compensation as in Fig. 3.28. Note that the first and last files are noise files (i.e. recording without firing a shot).

ably performed to zero offset, i.e. coincident source and receiver. Defining the sample values on trace i to be $a_i(t)$, the mean will be given by

$$A(t) = \frac{1}{N} \sum_{i=1}^{N} a_i(t) \qquad (3.3.5)$$

where N is the fold of stack. The process is illustrated by Fig. 3.39. This summation or stacking of traces is normalised in the above result by the factor $1/N$. This results in the signal having correct amplitude after the stack. However, this normalisation may not be optimum when the presence of noise is considered.

3.3.5.2 Signal to noise improvement

As a brief prelude to alternative normalisations, consider the signal to noise improvement attainable in a stack. As was discussed in Chapter 2, if each trace after moveout correction is thought of as the same signal plus Gaussian noise, the signal to noise improvement in the stacking process should be

$$(N)^{1/2}$$

This is the maximum improvement obtainable and the conditions to achieve this are:

1 The amplitude levels of each trace over the same time zone should be equal.
2 No bias (added constant) should exist on any traces.

3.3.5.3 Stack normalisation

As was mentioned earlier, a normalising scalar of $1/N$ when applied to a simple sum of sample values will result in an output trace containing a signal of correct amplitude. The word 'correct' is used in the sense that signal amplitudes will remain at the same level as on the unstacked traces. If, however, portions of traces are stacked which contain only noise, the RMS amplitude of the simple sum increases as

$$(N)^{1/2}$$

and a normalisation factor of

$$(N)^{-1/2}$$

is required to balance the amplitudes correctly.

In practice, this is recognised and accommodated by the use of a scalar

$$(N)^{P}$$

where p lies between -1 and -0.5. Other functions of N are also in use.

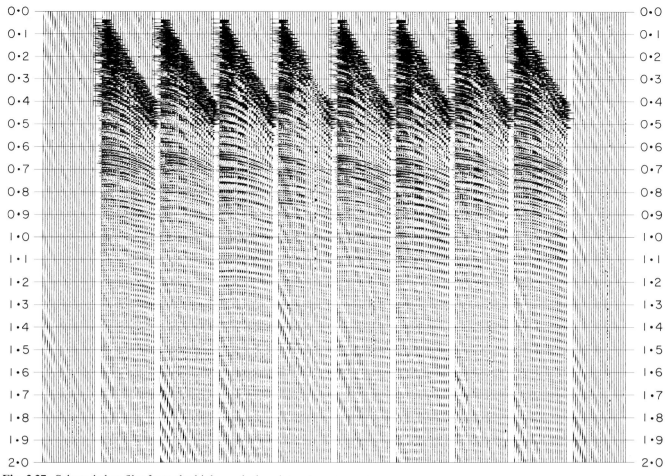

Fig. 3.37. Selected shot files from the high-resolution dataset
with no compensation as in Fig. 3.29.

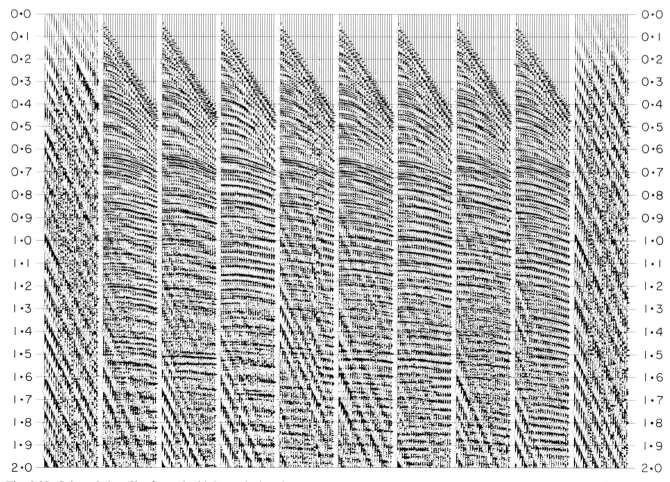

Fig. 3.38. Selected shot files from the high-resolution dataset
with a 400 msec window AGC as in Fig. 3.30.

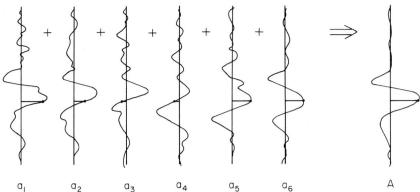

a_1 a_2 a_3 a_4 a_5 a_6 A   **Fig. 3.39.** The action of stacking.

In some circumstances it is desirable to replace (3.3.5) by a weighted stack defined as

$$A(t) = \left(\sum_{i=1}^{N} w_i \right)^{-1} \sum_{i=1}^{N} w_i \cdot a_i(t) \qquad (3.3.6)$$

where w_i is the weight to be used on the i th trace. This weight may be chosen by the geophysicist or according to some more sophisticated model of noise attenuation.

3.3.5.4 Effects of normalisation

On first appraisal, it may not seem important which value of normalisation scalar is chosen, as the fold of stack (N) should be constant along any marine line and the change in data amplitudes following stack can be corrected by a simple change in display level. Variation of N does, however, occur due to:

1 *Missed shots.* Whenever a shot is missed or traces are zeroed, the fold of stack over that region of line decreases. With well-acquired, high-fold data, the percentage decrease in N seen is usually quite small and no spatial amplitude variation can be detected. For low-fold data however, where many shots are lost, correct normalisation is required to prevent spatial variation of amplitude along the line.

2 *Offset muting.* In shallow data zones, muting (i.e. zeroing) of the offsets reduces the fold of stack dramatically. (See Section 3.3.7.2 for more details.) Careful normalisation is required to prevent further amplitude decay being introduced into the traces following stack.

In practice, use of a scalar close to

$$(N)^{-1/2}$$

will leave data at depth (high N) with amplitudes closer to those in the shallow.

As an example, consider the scalars used when 4 and 25 traces are live:

> $1/N$ normalisation. The ratio of scalars is
> $1/25 \div 1/4 = 0.16$
> $N^{-1/2}$ normalisation. The ratio of scalars is
> $1/5 \div 1/2 = 0.4$

When presentation of signal amplitudes for stratigraphic interpretation is important, normalisation using $p = -1$ should be performed.

Although the mean is by far the most common statistical estimator of signal amplitudes used in CMP stacking, other estimators can be used. The median and weighted mean can both be more effective at multiple attenuation, and are discussed later.

Fig. 3.40 shows a 'so-called' raw stack which corresponds to a first effort and is used to improve the seismic data processor's model of the particular dataset needed to improve the processing on a subsequent pass.

3.3.6 Stacking velocity derivation

3.3.6.1 Types of velocity

As an introduction to this section, it is worthwhile reviewing the types of velocity most commonly encountered.

Interval velocity (V_I) The velocity of a wave front through a single homogeneous layer.

Normal moveout velocity (V_{NMO}) The velocity which appears in the normal moveout equation and defined by the earth model used.

Root mean square velocity (V_{RMS}) The weighted root mean square of layer interval velocities, the weights being determined by the thickness of the layer as illustrated in Fig. 3.41. The inverse equation relating interval velocities to RMS velocity is known as the 'Dix equation'.

In addition to these, the following are often useful:

Average velocity (V_{AVE}) This is the mean velocity of the wavelet averaged over travel time. It has the same notation as Fig. 3.41.

$$V_{AVE} = \frac{1}{T} \sum_{i=1}^{N} V_i \cdot t_i \qquad (3.3.7)$$

where

$$T = \sum_{i=1}^{N} t_i$$

is the total travel time.

The average velocity is used to convert from travel time to depth.

Stacking velocity (V_{STK}) The stacking velocity is simply that velocity function which, when used for NMO cor-

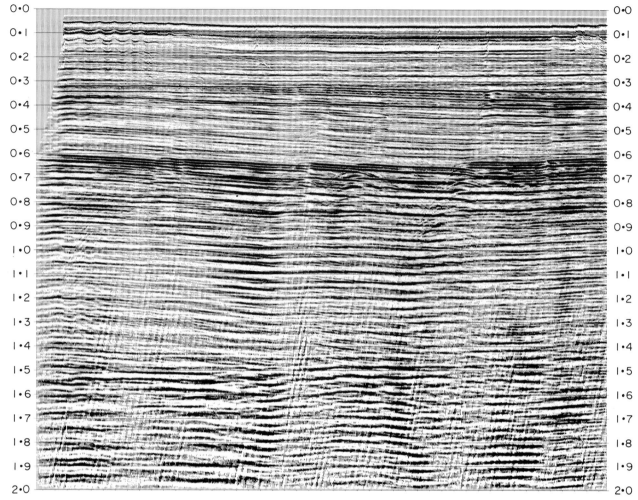

Fig. 3.40. A raw stack of the high-resolution dataset.

rection, produces the optimum stack. As such, it contains a great deal of subjectivity, quite apart from the fact that it is actually chosen from stacking velocity analyses. For example, it is quite common to force the stacking velocity to be unphysically high inside a salt diapir capped with a shale, to prevent multiples generated by the cap from ringing down through the salt. Hence, it may also have a cosmetic content.

There are a large number of other types of velocity in common usage some of which are, unfortunately, mis-

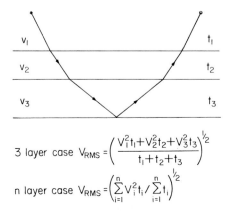

Fig. 3.41. The definition of RMS velocity, V_{RMS}.

takenly transposed, but the above list will temporarily suffice.

There are two general approaches to the extraction of velocity information from seismic data:

1 Reflection amplitudes can be related to acoustic velocity contrast across interfaces assuming an approximate knowledge of the density.

2 Detection of hyperbolic patterns within a CMP. The curvature of a hyperbola provides an estimate of the normal moveout velocity.

In general, reflection amplitudes provide information regarding local velocity variation whilst NMO hyperbolae define large scale velocity variation. In this section, only the latter will be considered.

The NMO equation derived as equation (3.3.3) for the single layer case, can be extended to a multi-layer earth quite simply. First rewrite (3.3.3) as

$$T_x^2 = T_0^2 + \frac{X^2}{V^2} \qquad (3.3.8)$$

where V is the interval velocity of the single layer. The extension of this for the multi-layer case is shown in Fig. 3.41 and results in the replacement of V in equation (3.3.8) by V_{RMS}, to give

$$T_x^2 = T_0^2 + \frac{X^2}{V_{RMS}^2} \qquad (3.3.9)$$

This derivation requires the approximation that the offset X is small compared to the depth of the interface with zero-offset time T_0, and is also known as the small offset approximation.

If the further complication of arbitrarily dipping layers is included, an analytical solution is highly complex even with a small offset approximation, although such solutions possess a rich and elegant structure as Hubral and Krey (1980) show. It is worthwhile here, however, to quote the solution for the case of single or multi-layer parallel dipping interfaces, as depicted in Fig. 3.42. The normal moveout velocity for this model is

$$\frac{V_{RMS}}{\cos(\theta)} \qquad (3.3.10)$$

Thus, the velocity required to correct any reflection for normal moveout will increase with the spatial dip of the interface. This result is important when estimating RMS velocities from NMO velocities.

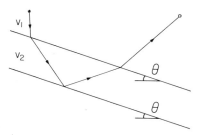

Fig. 3.42. For a conformal set of dipping layers, the moveout velocity is given by equation (3.3.10).

3.3.6.2 Stacking velocity analysis methods

From geometric considerations, the offset X for any trace is known. If a method of detecting primary reflections and tracking them across a CMP gather can be devised, T_x and T_0 can be obtained and equation (3.3.9) solved for V_{RMS}.

The object of such automatic analysis is to produce a stacking velocity function $V_{RMS}(t)$.

In theory, only arrival times on two traces are required to produce an estimate of V_{RMS}, but a more reliable estimate is achieved by using the whole CMP gather. In practice, a trial-and-error approach is tested in which many normal moveout velocities are used at specific increments to determine time shifts for each trace. The many types of velocity analyses currently in use vary only in the method of primary reflection detection.

3.3.6.2.1 Constant velocity stacks (CVS)

The simplest approach is to select a short panel of data of the order of 5 to 25 CMPs and produce a series of stacks using many time-invariant normal moveout velocities, ranging from the lowest velocity expected to the highest. Fig. 3.43 illustrates such an analysis. The

picking of a single optimum velocity profile is not easy, however, especially at depth where NMO velocities with a spread of 400 m/s produce comparably good results.

The weaknesses of the CVS analysis are its lack of resolution in velocity due to the often coarse increment for reasons of economy, and poor visual dynamic range.

3.3.6.2.2 The velocity spectrum

This is a commonly used piece of jargon which is used to describe a graphical display of some coherence measure, C say, as a function of NMO velocity and time. The word spectrum should not be taken to mean that the display involves some frequency domain attributes, it does not. The purpose of the coherence measure is to determine how well a hyperbolic trajectory of some trial curvature, dependent on T_0 and V_{NMO}, fits the data itself at a particular time, T.

A number of alternative coherence measures exist, each with their own particular advantages and disadvantages. Using an N-sample window of M traces after trial NMO correction, for example, the following can be defined:

Simple summation

$$C_{SUM} = \sum_{i=1}^{N} \left| \sum_{j=1}^{M} a_{ij} \right| \qquad (3.3.11)$$

Where a_{ij} is the amplitude of time sample i on trace j. This measure has the advantage of having excellent resolution in time, but poor resolution in velocity. The following two normalised measures are by far the most common.

Semblance

$$C_{SMB} = \frac{\sum_{i=1}^{N} \left[\sum_{j=1}^{M} a_{ij} \right]^2}{M \sum_{i=1}^{N} \sum_{j=1}^{M} a_{ij}^2} \qquad (3.3.12)$$

This measure, due to Neidell and Taner (1971), effects a good compromise and has good resolution in both time and velocity. Note that $0 < C_{SMB} < 1$, with perfect correlation corresponding to 1.

Considering only two channels, and letting

$$a_{i2} = k a_{i1}$$

for all i, where k is some constant, equation (3.3.12) reduces to

$$C_{SMB} = \frac{(1 + k)^2}{2(1 + k^2)} \leq 1$$

with equality only when $k = 1$. Hence the measure is sensitive to fluctuations of amplitude and thus produces a non-perfect correlation between traces identical apart from a scaling factor. This can be very useful in land data analysis.

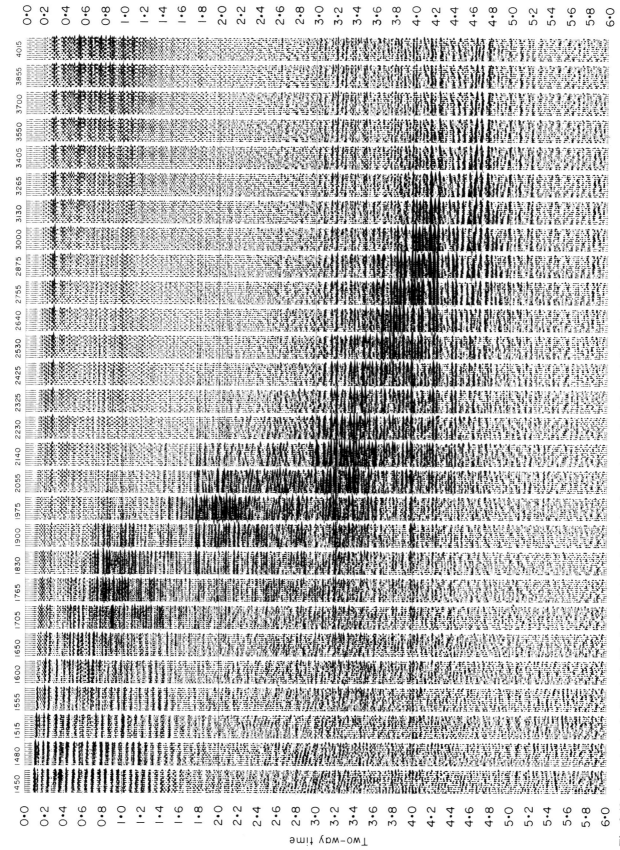

Fig. 3.43. A constant velocity stack. The horizontal axis is velocity, increasing to the right, and the vertical axis is time, increasing downwards.

Normalised cross-correlation

$$C_{NXC} = \frac{2 \sum\limits_{i=1}^{N} \sum\limits_{j,\,k=1\,(not\,j=k)}^{M} a_{ij} \cdot a_{ik}}{M(M-1)\sum\limits_{i=1}^{N} \sum\limits_{j=1}^{M} a_{ij}^2} \qquad (3.3.13)$$

This measure is normally defined for two channels as in equation (2.7.5), but has been extended for M channels here, compatible with its conventional usage in velocity analysis. It has a great deal in common with semblance, save for the fact that it is not sensitive to amplitude fluctuations between otherwise identical traces, as can easily be seen by using the two channel example calculated above for the semblance measure.

3.3.6.2.3 Combination displays

Velocity spectra are very convenient for good quality high volume data, but for more reliable work, various velocity analysis techniques are displayed together to give the geophysicist as much relevant information as possible in difficult areas. For example, stack panels, velocity spectra and optionally single CMP gathers after trial NMO are often combined. The stack panels are produced after NMO, usually according to a time variant velocity profile rather than a constant velocity.

The gather display is useful for checking the quality of offset muting and amplitude scaling along the line.

A further modern enhancement is the ability to do velocity analyses along specified horizons. These are particular useful for attaching geological significance to NMO velocities using various modelling techniques, and are usually done on data which has had a considerable amount of pre-processing.

3.3.6.3 Velocity analysis enhancement

The selection and pre-processing of traces used for velocity analysis can make interpretation considerably easier. Fig. 3.44a—f illustrate the effect on a contoured semblance display of changing certain key parameters.

Number of CMPs used Semblance values are commonly averaged over several CMPs. A reduction in random semblance peaks created by noise is achieved. Fig. 3.44b shows a velocity spectrum derived from 8 CMPs exhibiting fewer random contours than that derived from 2 CMPs (Fig. 3.44a). A continual improvement in interpretability with increasing number of CMPs cannot be expected, however, as geological dip results in a smearing of peaks in time. (A good pick at 1120 msec of 3100 m/s is poorly defined when 8 CMPs are used.)

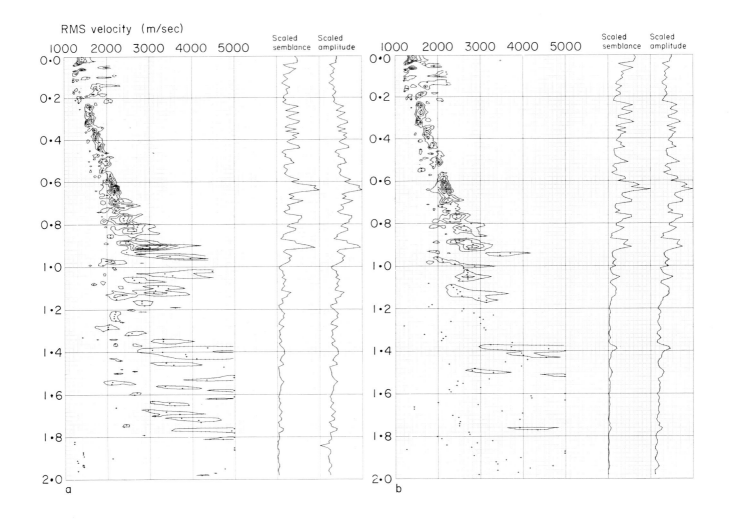

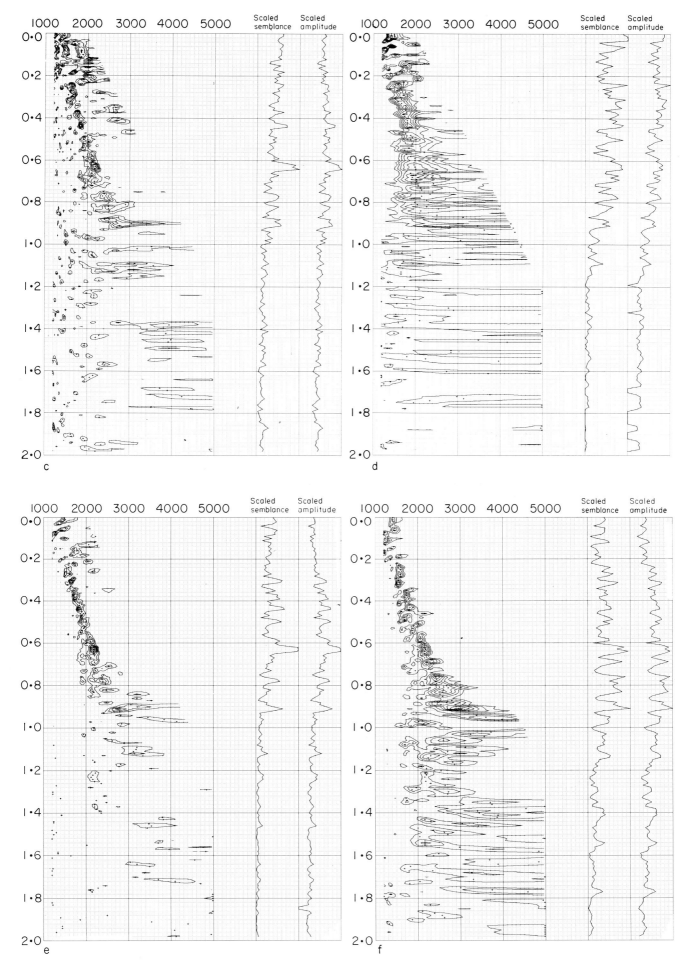

Fig. 3.44. Semblance spectra for the high-resolution dataset.
(a) Two adjacent CMPs used. (On opposite page.)
(b) Eight adjacent CMPs used. (On opposite page.)
(c) As (a), but every third trace used only.
(d) As (a), but only the near third of the cable used.

(e) As (a), but with low frequencies removed using a band-pass filter (80–240 Hz).
(f) As (a), but with high frequencies removed using a band-pass filter (12–80 Hz).

Number of traces used within a CMP When every third trace of a CMP is used for analysis as shown in Fig. 3.44c, random semblance peaks become more numerous and the display looks generally noisy. The shape of the contours nevertheless remains substantially the same. When the same data volume reduction is achieved by using the near third of the cable only, smearing of semblance peaks in velocity is pronounced and interpretation at depth becomes impossible, as shown by Fig. 3.44d.

Effective wavelet length The traces input to the analyses in Fig. 3.44e and 3.44f have been bandpassed to reject low and high frequencies respectively, in an attempt to produce effective wavelets of different duration. (Changes in wavelet length could also have been produced by suitable choices of deconvolution parameters.) As might be expected, temporal resolution of semblance peaks is better in Fig. 3.44e, but an improvement in velocity resolution is also noted. General interpretability however, is not so good as the signal-to-noise ratio has been degraded by rejection of low frequencies.

3.3.6.4 Checking routines

Stacking velocity interpretation is not complete after a velocity profile is picked at each analysis location. Restrictions placed on the general shape of each profile by geological considerations require check displays.

Interval velocity check The gradient of the picked stacking velocity profile is approximately related to the interval velocity of the rock present at any two-way time. Stacking velocity should increase rapidly within high velocity layers, and more gradually in slow sediments. A 'kick-out' in stacking velocity which implies an abnormally high interval velocity will probably need correction. Fig. 3.45 shows a very simple but valuable plot to assist in this.

3.3.6.5 Considerations for quality

A good stacking velocity interpretation involves the collation of many fragments of information and the making of value judgements as to their relative reliability. The following considerations can all influence the velocity 'picks' made:

Field files Inspection of refracted energy will confirm presence or absence of high velocity layers in the shallow section.

Geological input Well velocity logs may be available if the data is from a mature exploration area. However, the reader should be warned that the relationship between well log velocities and stacking velocities can be exceedingly tenuous.

Character of a coherent event Inspection of an inside trace or raw stack display can indicate whether the pick is primary reflected energy, diffraction, or multiply reflected energy. Dip, frequency content and structure across the section are all diagnostic of the validity of a given pick.

Quality of pick More weight should be given to highly coherent picks on events with little or very shallow dip when deciding upon a general velocity trend for a line. All the velocity analyses should be inspected before interpreting an individual one and the overall model should be built up from the few 'confidence-inspiring' ones. Picks which vary from this model wildly can then be treated as unlikely.

Consistency of profiles Individual analyses should be continually compared to analyses in the same neighbourhood. Changes in the general shape of profiles along a line will normally relate to similar geological trends.

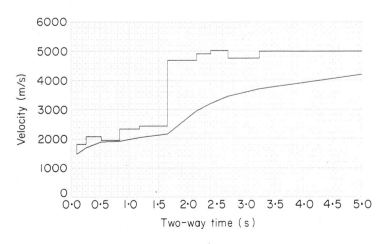

Fig. 3.45. Interval velocity bar chart produced from a RMS velocity function as a diagnostic.

Geological continuity Velocities should not be interpreted without regard for the general geological trends along the line. To ease the checking of velocity consistency with structure, contour maps of the picked velocity can be produced as an overlay for the raw stack and inconsistencies investigated.

3.3.7 Data preparation prior to stacking

Before the CMP gathers can be stacked, a certain amount of conditioning or pre-processing is required to ensure that an optimum stack results. The following topics need careful consideration.

3.3.7.1 Normal moveout correction

As was mentioned earlier, energy reflected from an interface will arrive at different receivers at times dependent on the offset of the receiver from the source. Corrections for these delays are made such that reflection arrival times on each trace are the same, and are equal to the two-way travel time that would be observed if source and receiver were coincident, i.e. at zero-offset. An undesirable side-effect of NMO correction is the deformation of the wavelet caused by NMO stretch. Normal moveout correction ensures that all theoretical reflection series spikes are shifted to the correct zero-offset time. In general, the amount of shift required will vary continuously from sample to sample down the trace as described by the NMO equation. In real data reflection series, spikes are replaced by wavelets, the samples of which are subjected to differential shifts producing non-linear distortion as illustrated by Fig. 3.46. This effect becomes pronounced when the rate of change of NMO (rather than the NMO itself) becomes large, and is generally seen at large offsets and shallow times. Very localised stretch is also sometimes seen deep in the data when a sudden 'kickout' in velocity is present.

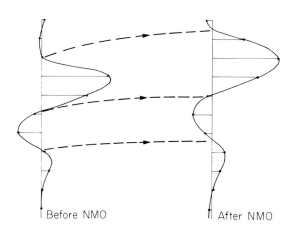

Fig. 3.46. The distortion inherent in moveout correction.

It can be seen from Fig. 3.46 that NMO correction cannot be restricted to whole sample shifts without intolerable distortion of wavelet shape and that some interpolation is therefore necessary. As was discussed in Section 2.9, the quality of this interpolator is important, especially where considerable stretch is involved.

NMO stretch places a minimum time restriction on the reliability of water bottom and near surface reflections for particular cables. For a conventional deep seismic cable with a short offset of 250 m, little interpretational value can be placed on reflections at two-way travel times less than two or three hundred milliseconds. Timing and wavelet shape will both be suspect. Parenthetically, this may also be true for other reasons, such as the dominance of refraction or direct arrival energy, and angle into the receiver array. For a high resolution cable with short offset of 50 m, reflections below 30 msec should generally be undistorted.

The stretching of a shallow water-bottom reflection is most severe when high velocity rock is present and the increase in stacking velocity with depth is rapid. The shape of the water bottom and near surface structure can be preserved if a false, high stacking velocity is used. Data should be muted to single fold in this zone to prevent attenuation of reflections during stack. This compromise approach will result in the correct spatial positioning of features but incorrect timing.

3.3.7.2 Offset muting

The exclusion of shallow long offset data from the stack is the normal practice for many reasons:

1 NMO stretch: as discussed above.
2 Wavelet distortion by the array: at high incidence angles, energy entering the receiver array will exhibit appreciable phase change across the hydrophones. Smearing of the composite pulse and loss of high frequencies occurs. Fig. 3.47 illustrates. These are known as directivity effects.

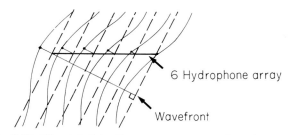

Fig. 3.47. Wavelet distortion across an array of receivers.

3 Dominance of non-reflected energy: direct arrivals, refracted energy and mode converted waves become dominant at high angles.
4 NMO assumptions: the approximation of NMO to a hyperbola begins to break down when offset approaches the same magnitude as reflector depth.
5 Complex ray paths: in structurally complicated areas, NMO can deviate strongly from hyperbolic and it may be beneficial for retention of high frequency information if only the near part of the cable is stacked.

In practice, a mute chosen to exclude the direct arrivals and above is a good first guess. At medium to large offsets a lag of three or four wavelet lengths should be allowed after the calculated direct arrival onset times, to allow the water-bottom reverberations, which arrive shortly afterwards, to decay. These multiple reflections will be relatively high amplitude due to the low absorption of energy in water, and closely spaced as they asymptotically approach the direct arrival time. A long offset mute is illustrated in Fig. 3.48.

For the near offsets, selection of a mute is more difficult and will often vary with water-bottom depth. If the water bottom is so shallow that a confusion of reflected, refracted, and direct arrival energy is seen on the near offset traces, there is no possibility of muting out unwanted energy whilst retaining primaries. Some traces from the inside of the cable must be kept live, but

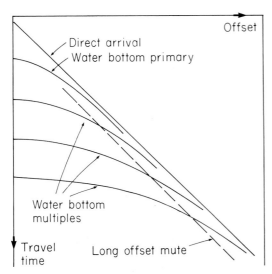

Fig. 3.48. An example of a long offset mute.

caution is necessary in interpreting this shallow data after stacking. A near offset mute is illustrated in Fig. 3.49.

In deep water, contamination of the inside traces with direct arrival and refracted energy does not occur and the mute may be left relatively open. Muting of the far offsets can also be less severe.

The reasons for muting have already been outlined, but whilst bearing these in mind, as little data as possible should be muted. Signal to noise will be improved and multiple attenuation enhanced, the greater the offset allowed into the stack.

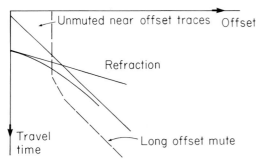

Fig. 3.49. An example of a near offset mute.

Finally, to illustrate the effects on real data, part of the high-resolution dataset is shown before and after muting in Fig. 3.50.

3.3.7.3 *Amplitude*

For a statistically optimum stack, data amplitude levels on input traces should be equal about any time scan. In reasonable data quality areas, well-acquired traces will be sufficiently equalised by the application of a normalisation scalar derived from a single large window. The window should be chosen such that the same reflections are included for all traces. This approach is not strictly valid when relative amplitude needs to be preserved, as scaling will be dependent on geology and noise conditions.

If normalisation of each trace is not performed,

however, incorrect channel recording levels or occasional noisy channels will result in a deterioration of stack quality. In very noisy data areas, consideration should be given to the application of a short sliding window (AGC) type normalisation, which will equalise mean amplitude variations both in time and offset. When reflection amplitudes of interest are comparable to the noise, this approach will not damage signal to noise, but will ensure that the effect of high amplitude coherent noise trains is reduced.

3.4 Data enhancement techniques

3.4.1 Temporal resolution

3.4.1.1 *Introduction*

A primary objective of seismic data processing is to recover the reflection series from a recorded trace. As was discussed earlier in Section 2.4.1, a trace can be represented by a convolutional model involving the reflection series e, corresponding to the earth's layering, as

$$s = w * e + n \qquad (2.4.3)$$

Ideally, the seismic wavelet could be deconvolved out of the trace s, to leave a spike series whose timing and amplitudes define the position of, and impedance contrast across, lithological boundaries. The presence of noise and uncertainty in wavelet shape however, require that the best that can be achieved is to replace the wavelet w with a more desirable one, w', in some sense.

The selection of w' is determined by the bandwidth of w when compared to that of the noise n. To prevent the trace becoming dominated by the generally wide band noise, a wavelet of restricted bandwidth is usually chosen. Even when complete spectral whitening is attempted, inaccuracy in the estimation of w will result in an actual wavelet of finite length. The choice of phase for w' is usually a simple matter. Inevitably either minimum or zero phase is selected, each one possessing certain merits.

3.4.1.2 *Criteria for selection of a desired wavelet*

3.4.1.2.1 Introduction

The amplitude spectrum of the existing wavelet is unlikely to be a smooth, optimum width function. It has already been seen in Section 3.3.3, that w is produced from the convolution of many physical effects and hence will have a notchy and generally narrow amplitude spectrum. In addition, the phase spectrum will not be purely minimum, but will contain mixed phase components. The wavelet w, therefore, is usually an extended one, with many peaks and troughs, and is difficult for the interpreter to associate with a single reflecting boundary. A more desirable wavelet would be one with a smooth, broader amplitude spectrum and of either minimum or zero phase.

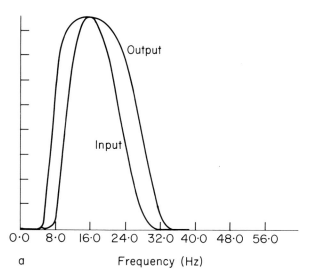

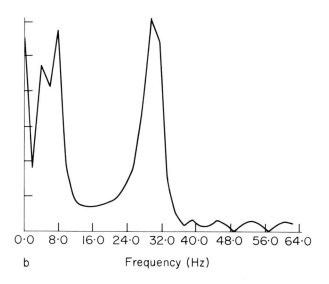

Fig. 3.54. Undesirable operator spectrum resulting from an injudicious choice of output wavelet spectrum. Strong amplification by the operator of narrow frequency zones is to be avoided.

only an estimate based on recorded traces of one kind or another, polluted by noise. The amplitude spectrum should be checked for plausibility and detailed time sample checking should be done if a wavelet or response has been digitised from some graphical display. Truncation and tapering is likely to be advisable especially at the end of an extracted or recorded signature. The time-zero position of a signature may also need adjustment.

The desired output It is more instructive to define the desired output as phase and amplitude spectra rather than as a time series. The choice of amplitude spectrum is the hardest selection to be made and is based on a balance between resolution improvement and signal to noise (S/N) degradation. As the noise spectrum is broader than the signal spectrum, broadening the signal spectrum amplifies frequencies with poorer S/N and reduces overall S/N ratio. S/N estimates are notoriously difficult to make, and degradation is unpredictable for a specific filter. Desired output spectra are best chosen by inspecting the operator spectrum after an initial design stage, and re-iterating the design with a modified desired output if extreme amplification of any frequencies is apparent (see Fig. 3.54). Operator spectra exhibiting narrow peaks are especially to be avoided as any inaccuracy in the input spectrum invariably results in strong 'ringing' or sinusoidal presence in the filtered data trace. Note that this problem does not manifest itself in operator design diagnostics. Undesirable ringing is also introduced by distortion of the otherwise comparatively flat noise spectrum.

As a general rule, highly variable operator spectra are to be avoided.

The choice of the desired output phase spectrum is a much easier decision to make. It is chosen almost invariably as either minimum or zero phase. It is generally thought best to keep to minimum phase until the

final stages after stack when conversion to zero phase can be performed, although there are alternative schools of thought.

Operator length There is no theoretical minimum or maximum. The result, by definition, will be optimal in a least squares sense for any particular length. A poor result will be obtained if the length is too short. If the operator is too long, little additional improvement results although the computer cost of using the operator increases rapidly. In addition, departures from the convolutional model due to real earth effects will be exacerbated. The ideal operator should look like a dinosaur: thick in the middle and thin at each end. Figs. 3.55 and 3.56b indicate an acceptable operator and one which is too short. A good first choice is the sum of the input and output wavelet lengths.

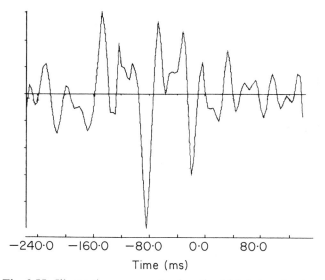

Fig. 3.55. Illustrating an operator length which is too short. The operator must have small amplitudes at both ends.

Operator time-zero position This issue specifically relates to the relative amounts of anticipation and memory components in the filter as discussed in Section 2.6.6. If insufficient anticipation component length is allowed, the performance of the filter may be significantly degraded. Most seismic software packages allow this to be computed automatically although the computation is expensive. If this facility is not available, the following simple rules may be of use.

1 If both input and output are minimum phase, no anticipation component is required.

2 If input and output are both zero phase, time-zero should be in the middle of the operator.

3 If either input or output contain mixed phase components, start with a first iteration of a fairly long filter with time-zero in the middle. Inspect the resulting filter and estimate new start and end times to be at positions where the filter amplitude has effectively disappeared, and recompute the filter accordingly, leaving the time-zero at the original first-iteration sample value. The procedure is fairly robust as the performance of such a filter varies only slightly around the optimum time-zero position. See Fig. 3.56.

The near-field signature, *p* Techniques for inverting the source signature deterministically have undergone considerable development over the years, to the point where the following options are open to the processing geophysicist.

(*a*) *The near-field signature (i.e. 1–2 metres from the source), has been measured directly for each shot.* The earliest successful attempts at doing this involved the use of a single element explosive source called Maxi-pulse.* Unfortunately, like all single element sources in routine marine acquisition, it was superseded by the power of multi-element sources (arrays) like air-guns, water-guns and steam jets. However, an advantage which single element source recording possessed until quite recently, was the fact that the near-field recording could be used directly as a far-field signature, if the appropriate geometric effects were taken into consideration. One of the disadvantages of single element sources was that they tended to vary sufficiently from shot to shot so that a recording and the corresponding Wiener deconvolution operator was necessary at every shot.

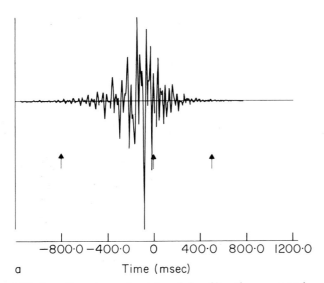

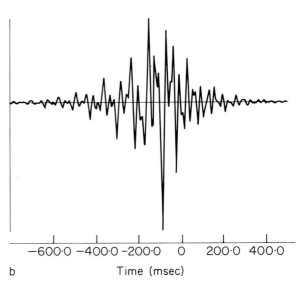

a   Time (msec)   b   Time (msec)

Fig. 3.56. Iterative scheme for determining filter time-zero and length. Positions of time-zero and selected start and end times used for iteration (b) are indicated on initial filter (a).

White light percentage This parameter is used whenever spectral equalisation of low amplitude frequencies is performed to prevent extreme amplification of noise, or very high or low frequencies outside the signal band (bandwidth of good *S/N*). It is normally specified as a percentage of the zero-lag auto-correlation value and generally lies in the range 0.1–5.0 per cent. See Section 2.8.4 for more details.

3.4.1.5.2 Source component inversion

The properties of convolution enable the inversion of individual components of equation (3.4.1) to be considered separately.

The advent of multi-element sources, although providing many benefits, caused considerable processing problems. The main one is that the near-field recording cannot be used as the far-field signature because of directivity effects induced by areal arrays with dimensions of several tens and sometimes hundreds of metres both inline and cross-line. An additional complicating factor is the effect of interaction, whereby each source element pressure field is modulated by the aggregate of the others, thus affecting the near-field recording. In this case, only one of two things can be done.

In some circumstances, a far-field recording is available for each shot of the multi-element source, suffi-

* Trade mark of Western Geophysical.

ciently far away (at least 50 metres) and directly below the array. Such recordings must be done in deep water to avoid spurious echoes from the seabed. Consequently, they can only be done routinely during a deep water survey. This brings in the additional problem that the deep-towed hydrophones used to do the recording tend to 'fish-tail' from side to side and so include directivity effects. They are therefore a little unreliable and may have to be averaged over several adjacent shots. It is important to note that such a recording also includes the source ghost — near-field recordings are considered unaffected by the source ghost as the hydrophone is usually much closer to the source element than is the surface.

If near-field recordings are available for each source element, a number of sophisticated techniques can be used which allow the far-field signature to be predicted with arbitrarily good accuracy, including both the effects of directivity and interaction. The reader is referred to Ziolkowski *et al.* (1982), Parkes *et al.* (1984).

An example from data acquired as part of the Delft experiment (Ziolkowski, 1984a) is shown in Fig. 3.57. Fig. 3.57a shows the far-field signature as predicted using the method referred to above. For the purposes of

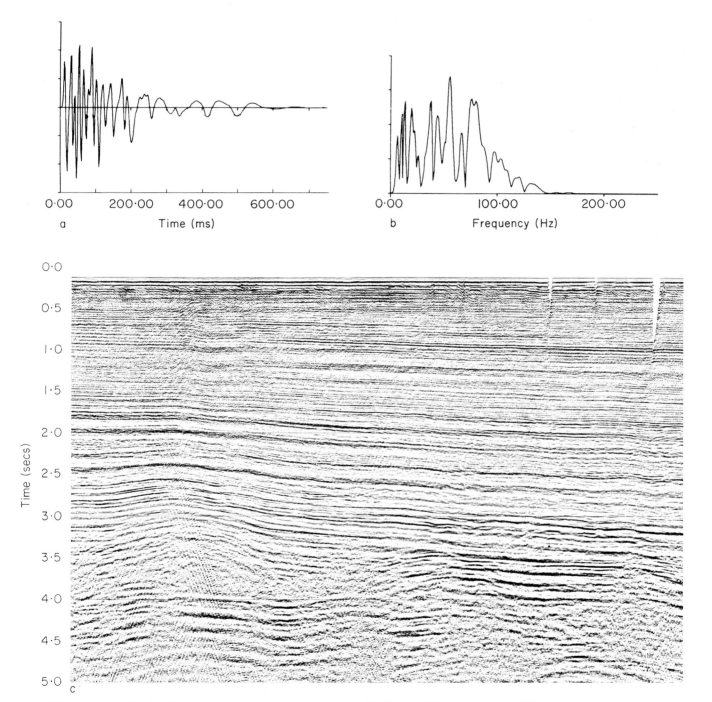

a   Time (ms)

b   Frequency (Hz)

c

Fig. 3.57. A case study of designature using a predicted far-field signature for an air-gun array using interaction according to Ziolkowski *et al.* (1982) and Parkes *et al.* (1984).
(a) The predicted far-field signature.

(b) The amplitude spectrum of (a).
(c) The dataset without designature.

(Figure continued over page.)

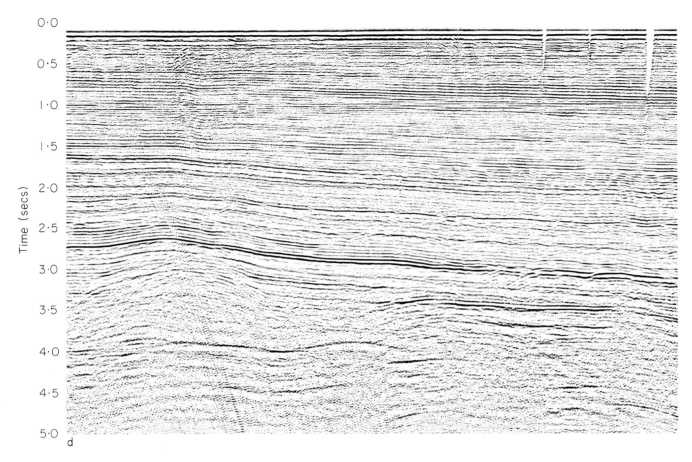

Fig. 3.57. Continued.
(d) The dataset after designature using the single Wiener filter designed to shape (a) into (c) and actually producing (d).

this experiment, a very poor quality source signature has been generated quite intentionally, by applying random time-shifts for each gun in a multi-element array. Fig. 3.57b shows the amplitude spectrum of 3.57a. Fig. 3.57c and d show the dataset before and after designature respectively, and vividly illustrates the power of this technique.

(b)   *The near-field, or far-field, signature has been measured under ideal conditions and the source is sufficiently reliable for departures from this archived signature to be insignificant.* This speaks for itself. It is certainly true that air-gun arrays deployed appropriately, have extremely consistent signatures, cf. Parkes *et al.* (1984). however, it is always possible that an element of an array will fail, changing the effective signature. Constant monitoring is therefore desirable to allow recomputation of the signature as and when necessary.

In spite of these caveats, marine sources do tend to be very consistent and the usage of a single signature is often an excellent working approximation, providing the sources are deployed in such a way as to minimise depth variations and maintain the relative geometry of the elements and the signature hydrophones. Many seismic processing centres capitalise on this fact to the extent of maintaining 'libraries' of standard signatures defined by array configuration, depth, filter settings and recording equipment. Fig. 3.58 illustrates far-field signatures from a variety of different guns.

Source and receiver ghosts, g_s and g_r. The ghost response, as introduced in Sections 2.3.7 and 2.4.4 is the *bête noire* of signature deconvolution. Removal of this response would at first seem to be a simple matter of designing an inverse filter and applying it to the data. This is not so for several reasons:

1   The amplitude spectrum contains a severe notch at a frequency determined by the source depth which if fully compensated for, will cause extreme amplification over a narrow frequency band.
2   Variations in source depth during the survey moves the notch.
3   Considerable variation in the ghost reflection coefficient is also possible during the survey depending on sea state and other factors, cf. Loveridge *et al.* (1984).

Taken together, these factors can dissuade the most ardent wavelet processor. It is usually better to rely on some less pathological spectral flattening. Some explicit correction can be achieved by:

1   Not attempting to broaden past the notch. This is done by specifying a desired output which decays considerably before the notch frequency. This is only viable if the source and receivers are sufficiently shallow to provide adequate bandwidth. Historically, this bandwidth restriction has been assumed unavoidable and surveys were planned with this frequency limit in mind. Currently, source arrays with components at mixed depths are being tested which alleviate the problem to some extent.

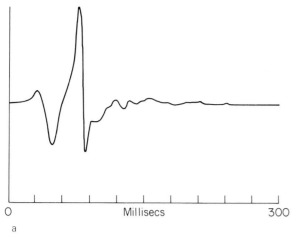

a

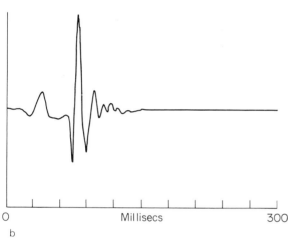

b

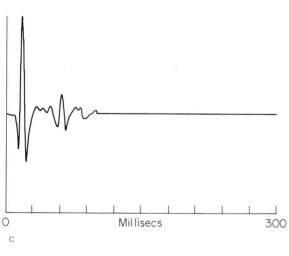

c

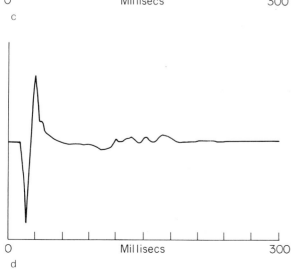

d

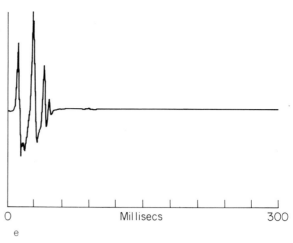

e

Fig. 3.58. Far-field signatures from a variety of guns.
(a) Starjet.
(b) Water-gun (400 cu in).
(c) Small air-gun (48.9 cu in array).
(d) Large air-gun (3640 cu in array).
(e) 30 Kjoule sparker.

2 Broadening past the notch but reducing the severity of notch depth and fluctuations by using a ghost 'wavelet' spread over a few samples centred on its expected time delay, for example,

$$(1, 0, 0, 0, -.9)$$

might be replaced by

$$(1, 0, 0, -.2, -.5, -.2)$$

The difference in the amplitude spectra can be seen in Fig. 3.59.

Note that angular variations in either signature or ghost are normally ignored, although a more detailed discussion assessing these effects is given by both Loveridge *et al.* (1984) and Ziolkowski (1984b).

Instrument response, *i* Although once quite popular, improvements in instrument technology now largely circumvent the need. Also, source signature recordings

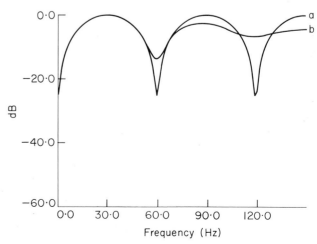

Fig. 3.59. (a) The amplitude spectrum of the ghost function $(1, 0, 0, 0, -.9)$.
(b) The amplitude spectrum of the ghost function $(1, 0, 0, -.2, -.5 -.2)$.

under survey instrument settings contain the response, again relieving the geophysicist of making explicit corrections. When done at all, only the phase spectrum of the instrument response is corrected. It does have considerable importance in a rather special environment, that of mixed land and marine acquisition, as geophones and hydrophones are typically out of phase by 90 degrees, as the former usually measures velocity whereas the latter measures pressure.

Earth filtering This is not explicitly mentioned in equation (3.4.1). Nevertheless, it is one of the most important effects to account for. In essence, it is desirable to accommodate attenuation effects (Q). Unfortunately, such effects are not convolutional. In practice, therefore, the wavelet is assumed to be constant within a window. Within this window, a correcting filter designed to be constant outside the signal band for stability, can be applied with a value of Q either supplied geologically or estimated by comparing the far-field signature of the source with some extracted wavelet from the data; a useful but controversial technique.

3.4.1.5.3 Wavelet estimation

Wavelet estimation must be resorted to when recordings are unreliable or simply not available, or when the effect of a geologically complex or unknown subsurface needs to be estimated. Insofar as such a thing as the wavelet can be defined, many different methods have arisen, common only in having unpronounceable names, and none acquiring dominance, although there is some seasonal fluctuation! All methods involve considerable averaging to improve the S/N ratio sufficiently to be applicable. A number of these methods will now be described.

Dominant horizon extraction In this technique, an average is taken of the water-bottom reflection or any high amplitude reflection from a simple interface. Such averaging is intended to remove geological effects which complicate the interface as well as to improve basic S/N. In addition, angle dependent effects can be catered for by considering only traces with small offset. The technique works best with deep, well consolidated water-bottoms with either no underlying horizons or only non-conformable ones. It is also wise to taper the extracted wavelet. In other cases it should be treated only as a source of comfort.

Spectral averaging Such techniques usually involve average auto-correlations of selected trace windows. To complete the puzzle, either minimum phase is assumed or it can be forced using an exponential taper as a pre-conditioner. The resulting wavelets are almost invariably smoothed to the point of being only indicative.

Sophisticated methods Such methods include such erudite concepts as minimum and maximum entropy,

parsimony, the complex cepstrum and others. They do not generally rely on assumptions about phase, but replace them with other assumptions such as sparse reflection series. There appear to be no general rules for predicting their usefulness, although they are not normally robust and either work well or fail miserably in practice. They all tend to perform magnificently on synthetic data.

3.4.1.6 Statistical deconvolution using Wiener filters

This is by far the most popular technique for spectral broadening due to its robust qualities and ease of use. Its effects are, to a certain extent, predictable, although test panels are always run to determine the best parameter choice. What constitutes the best choice is, unfortunately, highly subjective and most people would agree that the process does something nice to seismic data but nobody is quite sure what. The mathematical details were described in Section 2.8, but the assumptions should be stressed again. These are:
1 That the reflection series is random and white.
2 That noise is random and stationary.
3 The wavelet is minimum phase.
The effectiveness of the deconvolution will be loosely related to the suitability of these assumptions.

The S/N ratio of the data also has an influence on the performance of the deconvolution and determines the amount of spectral broadening to some degree. Traces with poor S/N produce high zero-lag auto-correlation coefficients, and thus produce a similar effect to that of using large amounts of white light. In good S/N areas, trace auto-correlations provide better estimates of the wavelet spectrum and deconvolution is more effective. Hence, large differences between test panels where varying degrees of wavelet compression are being com-

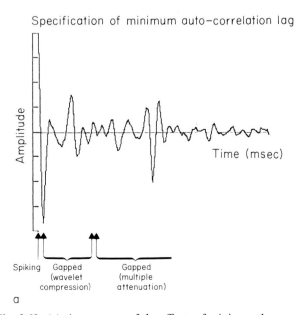

Fig. 3.60. (a) A summary of the effects of minimum lag specification for predictive deconvolution.

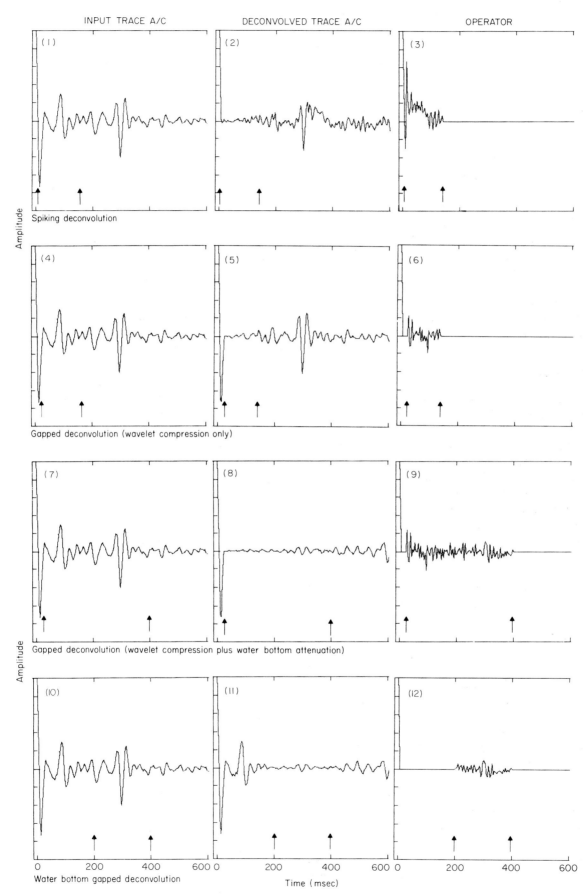

Fig. 3.60. (b) Minimum and maximum lag specification
(indicated by arrows) for predictive deconvolution.

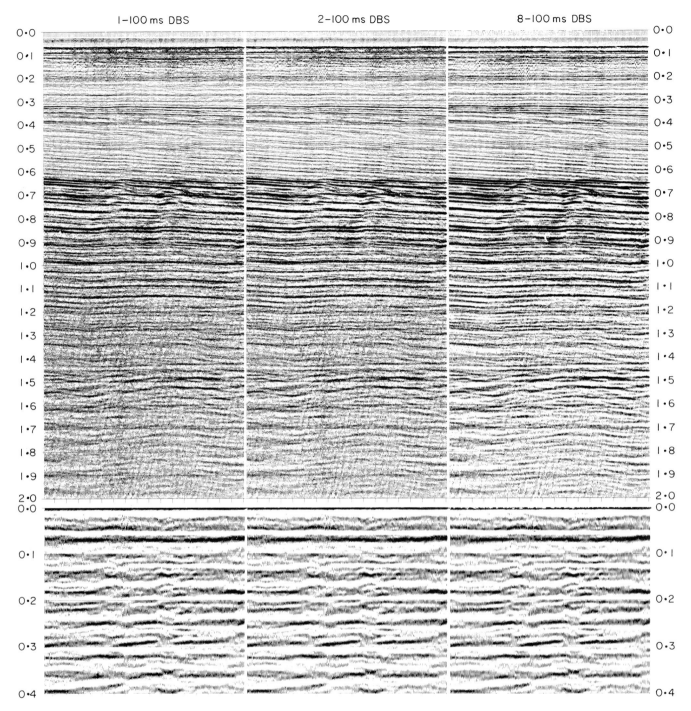

Fig. 3.61. A sequence of DBS (deconvolution before stack) panels. The high-resolution dataset has been used. One sided auto-correlation traces, computed from a selected window, are displayed beneath the data. Note the consistent water-bottom multiple at 50 msec period. Minimum and maximum predictive lags are specified above panels.

pared should only be expected for lines with good S/N. In poor areas, significant compression is simply not possible with this method. Note also that non-random features of the noise will result in a contribution to the other auto-correlation lags and contaminate the wavelet spectral estimate.

Data pre-conditioning is important in general. Anything which would result in the trace auto-correlation being unrepresentative of the wavelet auto-correlation should be avoided or corrected. For example:

1 Obvious noise trains such as refracted energy and so

on should be excluded from the window.

2 A window with high signal content should be selected.

3 Amplitude decay within the window should be compensated with a slowly varying gain curve.

4 The data should be pre-filtered to remove noisy frequencies within and outside of the signal band.

The parameter decisions to be made are minimum and maximum auto-correlation lag, auto-correlation window position and length and white light addition. These will now be considered in turn.

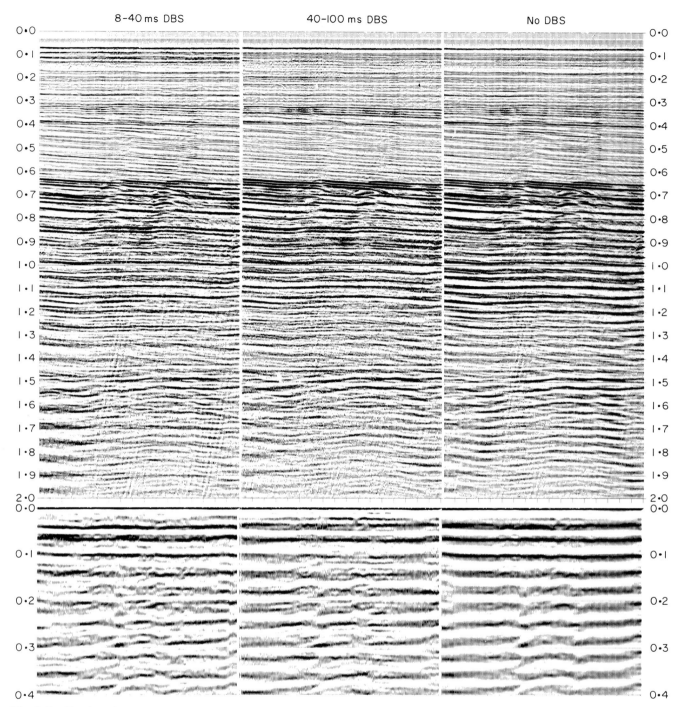

Fig. 3.61. Continued.

3.4.1.6.1 Intrinsic parameter choice

Minimum and maximum auto-correlation lag In essence, the Wiener–Levinson algorithm attempts to produce an output trace with zero auto-correlation values between the specified minimum and maximum lags.

The choice of minimum lag or gap as it is generally known, determines the amount of wavelet compression. When a one sample gap is specified, complete whitening of wavelet amplitude spectrum is attempted with only the auto-correlation zero-lag coefficient left untouched. This is commonly known as spiking deconvolution. When larger gaps are specified, the deconvolution is known as predictive deconvolution for reasons dis-

cussed in Section 2.8. For gaps longer than 2 or 3 auto-correlation function zero-crossings, little change in the wavelet amplitude spectrum occurs and the objective is short period multiple attenuation. Fig. 3.60a shows a typical auto-correlation and illustrates commonly used terminology. The gap can usually be specified in milliseconds or zero-crossings, the latter producing a more variable result from trace to trace.

The position of the maximum lag determines the amount of predictable amplitude removed. Generally, short maximum lags result in wavelet compression only, whereas longer maximum lags also perform multiple suppression (usually the water-bottom in marine data). The specification of minimum and maximum lags determines the active region of the operator as illustrated by Fig. 3.60b. Note that a gapped wavelet compressing deconvolution is sometimes simulated by adding a relatively large quantity of white light to a spiking deconvolution.

As might be inferred from this rather qualitative account, deconvolution tests are normally done as a series of panels to help determine optimum parameter choice, whatever that is! A sequence of such panels is shown in Fig. 3.61. The data has been stacked following deconvolution. S/N degradation at short gaps can be seen and this constitutes a key factor in parameter selection.

Auto-correlation window position and length Keeping in mind the consideration already mentioned regarding data preparation, the position of the window is generally determined by the primary zone of interest. The window length is a compromise between increasing statistical validity (long window) and the desire to retain stationarity (short window). A rule of thumb often quoted is ten times the maximum auto-correlation lag, but this is strongly affected by the S/N of the data and the density of reflections within the window. There is no advantage in defining a long window containing no reflections. Poor S/N can be ameliorated by using a multichannel approach whereby windows from consecutive traces are averaged.

Pre-stack, window position and sometimes length are varied according to offset from the source in an effort to include the same reflections within each window. Post-stack, interpolation between specified CMPs is performed to ensure that the zone of interest remains within the window.

White light White light functions exactly the same as in Wiener filter design, but as noise contributes considerably to the trace auto-correlation, it is not so important and is generally only considered for 1 or 2 sample gaps.

3.4.1.6.2 Global parameter considerations

Time variance — multiple window deconvolution So far, only a single operator per trace has been considered and time variance of the wavelet has been ignored. It is, however, common to use 2 or more operators determined from contiguous windows both pre- and post-stack. Windows are chosen to cover sequences of reflectors with the change of windows occurring in a 'quiet' zone. The positioning of each window is determined from observation of signal frequency changes associated with geology. Overlap of windows is permitted by most computer implementations and is especially necessary for far-offset traces pre-stack. Experience shows that regions with a deep target zone and thick overlying Tertiary deposits are suitable for 2 window deconvolution, the merge zone being above the base Tertiary or Cretaceous reflection. Regions with no Tertiary deposits and shallow target zones restrict the geophysicist practically to a single window. Regions which

consist mainly of sand/shale sequences often do not provide a well defined zone for window transition, and multiple windows can be arbitarily located as long as reasonable signal content is provided within each window.

In any multiple window filtering operation, the method of merging adjacent windows must, of course, be specified. There appears to be two methods in common use as shown in Fig. 3.62.

1 The first technique applies distinct filters in each window and then merges the window over user-specified merge zones. Outside the merge zones, the actual filtering applied is easy to evaluate in this method. The merge zones themselves are usually linear combinations of the contributing windows. Multiple window deconvolutions are normally applied using this merging technique.

2 The second technique uses a continual merging of windows so that it is very difficult to specify exactly what filtering has been applied at any one point in time. This technique is more commonly employed for the application of band-pass filters. See Fig. 3.62.

3.4.1.6.3 Conclusion

A number of common methods of increasing temporal resolution have been discussed. It should be noted, however, that many other processes can damage resolution and their effects need to be borne in mind in any such attempts. These include:

Band-pass filtering Unless carefully designed, this operation, apart from its primary purpose of attenuating those frequency bands of poor S/N, may degrade the signal content appreciably, resulting in wavelet temporal extension. It should be noted, however, that if the wavelet amplitude spectrum is particularly unbalanced, say through a richness in low frequency content, attenuation of these frequencies may well improve matters by reducing 'ring'.

Normal moveout and stack A certain amount of high frequency degradation is inevitable due to incorrect NMO application, NMO stretch, non-hyperbolic moveout and poor interpolators, especially at larger offsets. If higher resolution is desired, it can be achieved at the expense of increasing noise and multiple contamination by stacking using the near offsets only, say the first 6 or 12.

Trace summing and mixing Vertical and horizontal trace summation, long source or receiver array simulation, and some post-stack coherency enhancement techniques generally improve signal coherency at the expense of resolution.

Coherent noise attenuation, especially in the $f - k$ domain should not generally degrade resolution of primary data outside of the reject zones, although the caveats of Section 2.10 should be borne in mind.

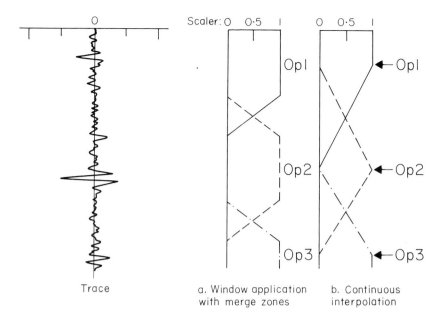

Fig. 3.62. Illustrating two methods of merging windows. Method (a) is commonly used for deconvolution filter application, whereas method (b) is used commonly for filtering. These somewhat *ad hoc* methods form the standard approximation to continuous time-variancy.

Display An often underestimated source of degraded resolution. It may sound trivial to say that if the plot is poor, the resolution is also, but it is a most important factor and is considered in detail shortly.

3.4.2 Random noise attenuation

3.4.2.1 Introduction

The mechanisms for the generation of essentially random noise have already been discussed in Section 3.3.3.1 and can be categorised into three types according to their attenuation properties.

3.4.2.1.1 Constant background noise

This is present effectively throughout the data and includes cable noise and instrument noise. It occurs in all surveys, and methods of attenuating it are included in all standard processing sequences. By far the most effective of these is the CMP stack, where a reduction in amplitude of $(N)^{1/2}$, where N is the fold of stack, is realisable in ideal conditions as stated in Section 3.3.5, although rarely if ever achieved in practice. Other optional techniques divide into single and multichannel methods.

Single channel band-pass filtering Filtering out noise from the data using a band-pass filter with suitable low and high cut-off points is commonly done late in the processing sequence when the useable signal bandwidth is better understood. This strategy makes the reasonable assumption that outside some signal band, noise is so dominant compared with the decaying signal content that elimination of these frequencies will reduce overall noise content whilst preserving wavelet resolution. Hence, the objective of band-pass filtering is exactly opposite to that of much deconvolution, which is attempting to expand the bandwidth. The choice of the

cut-off points is highly subjective, and it can be argued that such usage as a final processing step is simply to correct the inadequacies of preceding deconvolution steps. It is to be hoped that future developments in deterministic deconvolution techniques will allow post-stack wavelet deconvolution and band-pass filtering to be combined into one process which will take into account signal-to-noise spectral estimates.

Filter parameters are usually decided in exactly the same way as deconvolution parameters, on the basis of test panels like Figs. 3.63 and 3.64. Fig. 3.63 shows part of a suite of largely distinct frequency bands whose purpose is to give the geophysicist a visual idea of the signal band, and Fig. 3.64 is a series of practical filter panels.

The basic requirement as stated earlier is to retain enough signal bandwidth in order to have a reasonably interpretable wavelet. Hence the filter should have a pass band of 2–3 octaves centred over the dominant signal frequencies.

The gradual movement of the signal band to lower frequencies at later reflection times naturally forces band-pass filtering to be time-variant for maximum effectiveness. Again, this is achieved in an essentially identical manner to that of deconvolution as described in Section 3.4.1.6.2. Changing geology similarly forces spatial variation. Ideally, the same horizon should be filtered in the same way, but the spatial and temporal aspects of change sometimes conflict and this is simply not possible, requiring yet another trade-off.

Filtering is usually performed with a zero-phase filter even if the wavelet itself is minimum phase. Theoretically, this is weak as explained in Section 2.6, particularly if a minimum-phase deconvolution is to follow. The reason for this practice probably lies in a desire not to appreciably change the timing of wavelet peaks and troughs, especially immediately prior to final display. Distortion of the wavelet shape should in any case be small, as hopefully little attenuation of signal frequencies occurs.

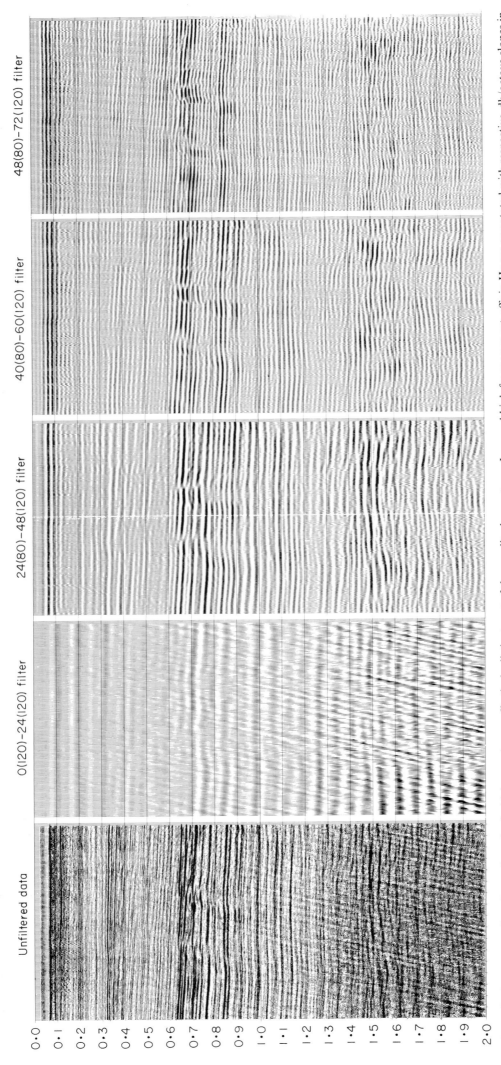

Fig. 3.63. A sequence of stack panels each of which contains an effectively distinct part of the amplitude spectrum. Low and high frequency cut-offs in Hz are annotated with respective db/octave slopes in parentheses. No amplitude equalisation is performed. Display of similarly narrow frequency band panels would routinely be continued up to the Nyquist frequency (256 Hz, in this case). Such a plot gives the geophysicist an excellent empirical feel for the signal band or bandwidth of good S/N. The high-resolution dataset has been used.

Multichannel techniques Band-pass filtering relies on being able to separate out signal frequencies from those containing noise. Multichannel techniques exploit the absence of trace to trace correlation in the noise. The CMP stack is of course a multichannel method. 2-D techniques such as array simulation, to be discussed shortly, and migration as discussed in Chapter 4, do not have noise reduction as their primary objective but usually achieve it as a 'fortunate' side effect. Often however, what was originally random noise prior to these processes, becomes organised into various characteristic patterns afterwards. The confused interpreter may well prefer the original higher level of random noise, given the choice.

Straight trace sums, mixes or stacks are additional methods of improving S/N ratios although, again, undesirable organization of random noise will take place. More sophisticated methods of noise reduction collectively known as coherency filters have been developed which improve signal within specific dip bands, but again the change in spectral character of background noise may be undesirable. The application of such processes is usually delayed to a very late stage as a 'cosmetic' step.

3.4.2.1.2 Noise bursts

These are generally present only in a small subset of the data and include such things as a bad recording channel and 'bird' (cable depth controller) noise. Such bursts should be detected at a quality control stage immediately after demultiplexing and treated as soon as possible. Techniques for their reduction include various gaining methods if noise burst amplitudes are not too high. Very high-amplitude bursts can be detected statistically and treated automatically. For such bursts, zeroing perhaps followed by reconstruction through spatial interpolation may be the best solution.

On marine data, such bursts usually occur consistently, for example, on a single bad channel or all channels of a weak shot, and editing is simple. On land data, noise bursts occur much less predictably and editing is often very difficult. One method used on land data to reduce the impact of such pollution automatically is that of diversity scaling, whereby traces are scaled inversely according to their effective noise content.

3.4.2.1.3 Spikes

These affect only a single sample or a few adjacent samples. They are extreme amplitudes and are generated by faulty equipment. Unfortunately, they are quite common and often occur as a spike before the instrument field filter is applied and so manifest themselves as a very large amplitude copy of the filter response over a few samples in the data. If the amplitude is not too large, this filter response can be very difficult to automatically distinguish from a genuine reflection.

Many corrective techniques have been devised for detecting and correcting such problems, based on characteristics like the amplitude level of individual samples, the average amplitude of small windows, and the rate of change of amplitude. The sophistication of any particular algorithm is judged by its ability to distinguish between anomalies and reflections, and very often they are programmed on a one-off basis for each type of occurrence.

3.4.3 Attenuation of unwanted signal

3.4.3.1 Introduction

Before discussing methods of attenuation, it is necessary to consider ways of recognising the presence of unwanted signal. This is generally a straightforward matter for coherent noise because of:
1 Linear moveout either across shot files or across the stacked section itself.
2 The frequency band is often low or limited.
3 The dip is often high and/or opposite to the primary reflection hyperbolae pre-stack, and bears no relationship to geological dip post-stack.

However, the identification of multiples is usually much more problematic because:
1 The moveout across the cable is hyperbolic or at least a close facsimile thereof.
2 The wavelet shape is similar to that of the primary reflections.
3 Continuous and highly plausible horizons remain after stack.

The following characteristics of multiples do build up confidence in an assessment of events of dubious origin although no single one may prove conclusive.

3.4.3.1.1 Periodicity

This is the major criterion for analysis. Multiples arrive at approximately a fixed period behind a 'generating primary' at any CMP. Any two high-reflection coefficient interfaces (including and especially the waterbottom, the free surface and the generating primary), will form the significant boundaries of a multiple reverberation, (cf. Fig. 3.16). The two-way time between the two reflectors on the stack section gives the lag following the generating primary at which the multiple is to be expected. The multiple should track rigorously at this lag across the section unless there are significant lateral variations in velocity. Amplitude variation should also imitate that of the generating primary. Second and higher order multiples may also be present assisting the identification. Simple multiples from the free surface and the generating primary reflection are easy to identify as dip is double that of the primary. In comparison, peg-leg multiples generated by shallow interfaces are much harder.

A pre-stack common offset gather shows multiples more clearly as stacking attenuates multiples because of their differential moveout, although the periodicity will be less well defined unless the common offset gather is corrected to zero-offset. Auto-correlation is often used to analyse periodicity.

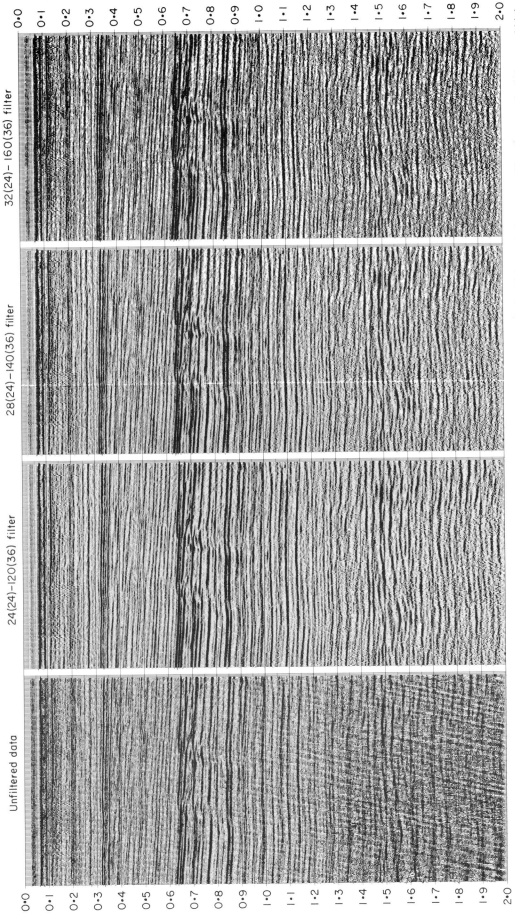

Fig. 3.64. A sequence of broad band filter tests displayed in a similar form to the DBS tests of Fig. 3.61. The dataset is the high-resolution one and has been stacked. Low and high frequency cut-offs in Hz are annotated with respective db/octave slopes in parentheses. Amplitude equalisation by means of AGC has been performed. Panels continue below.

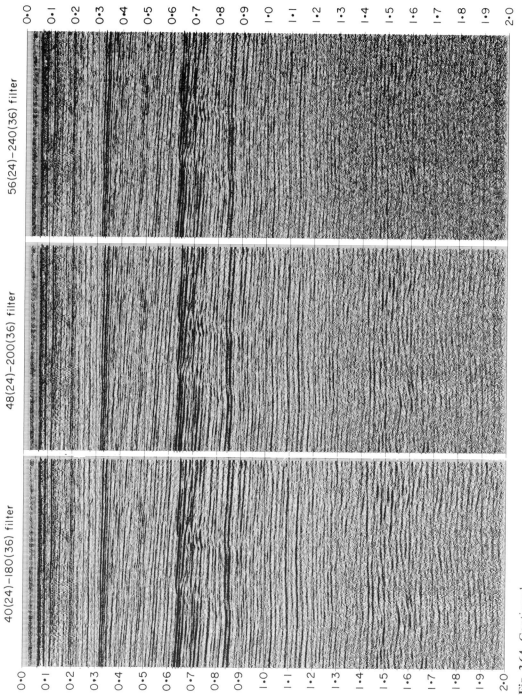

Fig. 3.64. Continued.

3.4.3.1.2 Low stacking velocity

The normal trend of increasing velocity with depth causes multiples to have a lower stacking velocity than primary energy at the same time. Consequently, multiple hyperbolae exhibit more moveout on uncorrected CMP gathers than do primaries; hence the term differential moveout. If velocity does not increase with depth, problems abound!

3.4.3.1.3 High frequency content pre-stack

Slower travelling multiple energy has a shorter travel path than primary energy and subsequently less high frequency is absorbed. This may be somewhat offset by the larger angle of incidence into the receiver array and emergence out of the source array. After stack, frequency content will be largely determined by primary/multiple NMO differential, and multiples will generally look low frequency due to the high frequencies stacking out of phase.

Relative amplitude is a particularly unreliable indicator of reflection validity. As first and higher order multiples do not decay in amplitude with the same spreading loss as contemporary primary energy (cf. Section 3.3.3.2), it is quite possible for strong multiples to be present in a stack section whilst the generating primary is hardly visible. As a consequence, although many orders of multiple may be recognised in this situation (usually shallow data beneath a very hard water-bottom), the first multiple is sometimes interpreted as the generating primary.

3.4.3.2 General attenuation techniques

Assuming that the problem is low velocity multiples or coherent noise, virtually any pre-stack mixing or summing results in degradation relative to the primary energy. The stack itself, of course, is a powerful attenuator.

The option exists to correct for differential NMO prior to the mixing of traces, in order that primary data adds constructively. If this is included, attenuation of multiples and noise travelling down the cable will be lessened. A single representative velocity function is adequate for this purpose.

Mixing of data pre-stack may be thought of in terms of lengthening the source or receiver arrays used in the field. Long arrays have a more directional response to incoming or emanating waves, vertically travelling waves being preferentially treated, as shown in Fig. 3.65. Multiple energy and source generated noise having travel paths at high angles to the vertical will consequently be discriminated against. Array directivity has been much studied over recent years, and physically long and wide acquisition arrays are commonly deployed in regions with known multiple or noise problems. The simulation of long source arrays by means of appropriate mixing during processing has become similarly popular. If linearity of the recording instruments is

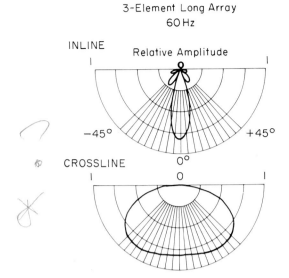

Fig. 3.65. A typical in-line and cross-line directivity plot for a long array of some 200 m length. Only 60 Hz is treated. Other frequencies are broadly similar with more directivity at higher frequencies. Note the very considerable attenuation in the inline direction for small angles away from the vertical.

assumed, the simulation should result in a very similar section. That they are not exactly equivalent is the subject of a paper by Ursin (1978).

3.4.3.2.1 Simulated long arrays

Array simulation is performed either in the common shot domain for receiver arrays or in the common receiver position domain for source arrays. For conventional shooting configurations, three, five, or more rarely seven, traces are mixed and the output used to replace the original central trace. The length of the simulated array is more often specified than the number of traces involved, and usually a weighting function is applied such that the extremes of the array are scaled downwards. The weighting function is either simply a linear taper or occasionally a spatial filter carefully designed to attenuate certain spatial wavelengths. In this latter case, many traces may be involved, some of which could be reversed in polarity.

The ease with which a desired array length and plausible weighting function can be chosen, the processing run submitted and the results obtained is, unfortunately, deceptive. In particular when simulating source arrays only certain lengths are meaningful and the geophysicist should always make certain that he or she knows which traces will, in fact, be mixed. A surface stacking chart rather than a subsurface stacking chart, as was shown in Fig. 3.12, is used as an aid to understanding as the common receiver position domain is shown more clearly. Fig. 3.66 indicates which traces are mixed to simulate a 50 m source array. The acquisition configuration was as for the subsurface stacking chart, Fig. 3.12. Note that although the shot pull-up is 12.5 m, only three traces are mixed rather than five, and that these are from alternate shots. Also, extreme traces on the

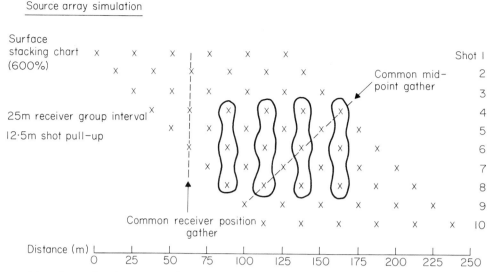

Fig. 3.66. Surface stacking chart: showing traces mixed to simulate a 50 m source array for shot gather 6.

cable will be incompletely mixed and should be either amplitude normalised accordingly, or rejected.

Rather than enhancing vertically travelling energy as simple source simulation does, small bulk time shifts, linearly dependent on position within the array, can be applied, which have the effect of rotating the plane of the extended array. This may be advantageous in some areas where the vertical is not the preferred direction for recording primary energy. This technique rejoices in the name of 'beam steering'.

Note that application of differential NMO corrections instead of linear time shifts 'focusses' energy into the receiver array. This is not a simulation of a physically realisable technique.

Source simulation often produces good results at depth in the stack section where primary NMO is small, but degrades the high frequency content of shallow primary data (large moveout/high angle ray paths) to an unacceptable degree. An alternative method of retaining shallow high frequencies, again physically unrealisable, is to vary the simulated array length with time such that mixing in the shallow is only over three traces with severely tapering weights, or none at all.

Finally, a word of caution is in order. If a long source or receiver array alone is constructed, the stack section will exhibit line-ups of noise predominantly in one direction owing to the inbalance of the two directivity functions. Mild mixing in both common shot and common receiver position domains to achieve comparable array lengths is preferable.

The problem of degradation of primary high frequencies when attenuating coherent noise is overcome by most dip filtering techniques. Naturally, other undesirable characteristics of these techniques arise instead.

3.4.3.2.2 2-D dip filters

For these processes, a range of dips is specified, usually in msec per trace, within which coherent energy is attenuated. In some computer algorithms, multiple dip

ranges may be specified. In an analogous manner to the 1-D case, 2-D filters may be applied either convolutionally in the $x - t$ domain or multiplicatively in the $f - k$ domain. The $x - t$ approach becomes expensive when more than a small number of adjacent traces are used, and with the advent of array processors (cf. Chapter 1) $f - k$ filtering techniques have become popular.

3.4.3.2.3 F − k domain filtering

Although $f - k$ filtering is a very powerful and commonly used processing tool, it must be used carefully to avoid possible processing artefacts. The following points are important.

Dip range selection The act of choosing the shallow dip edge of the attenuation fan is invariably a compromise between allowing some noise to remain and damaging primary data. This results quite simply from the fact that the signal and noise areas in the $f - k$ domain overlap. The desired dip is chosen most safely by looking for the highest primary dip and slightly adding to this. After stack, this means inspecting the geological features; before stack, the highest remaining dips of primary hyperbolae after the application of the stacking mute. In geologically complex areas, where moveout is contorted, this is no simple matter!

Pre-stack, a compromise may have to be made where the outer traces of shallow primary data are attenuated in order to improve noise reduction in deep target zones. Time variant dip specification, not commonly available at present, would avoid this compromise.

The high dip edge of the fan is easier to select as no primary data is in danger and it may be made very steep if required.

Spatial aliassing The choice of high dip is relatively unimportant only if the scaling of the $f - k$ domain wedge is terminated at the spatial Nyquist, as in Fig.

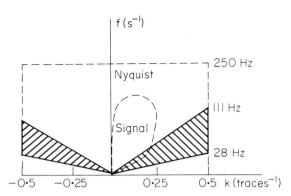

Fig. 3.67. A schematic of the attenuation zones for the $f - k$ coherency filter used in Fig. 3.68.

3.67. In this case, frequencies higher than 28 Hz would not normally be attenuated for this highest dip as they are spatially aliassed. Some computer algorithms may allow filtering to take place past the spatial Nyquist so as to wrap-round into the opposite quadrant. The risk of degrading primary data is high in this case as was discussed in Section 2.10, particularly post-stack when data has both positive and negative dips. The presence of high frequency remnants in shot gathers is generally not a serious problem as stacking attenuates such noise effectively.

$F - k$ domain tapers Just as a steep slope in the amplitude spectral representation of a 1-D filter produces ringing, insufficient tapering between the filtered and unfiltered regions of the $f - k$ domain introduces coherencies into the filtered data at dips corresponding to the fan edges. Note that the slope of the taper is affected by the severity of the attenuation scalar.

A large taper zone, of course, reduces the ability of the filter to discriminate between primary and coherent noise data of similar dips. Various ways of overcoming this problem have been proposed including:

1 Defining the edge of the reject region as an arbitrary polygon or curve such that different frequencies cut-off at different dips rather than the same dip.

2 Automatically threading the edge of the cut-off region through low amplitude channels of the $f - k$ domain, assuming that such channels are suitably situated. (The slope of the $f - k$ taper is unimportant if applied in a region of the $f - k$ domain where sample values are small.)

The nature of the taper, of course, affects the $x - t$ impulse response of the filter. It is instructive to check this response if possible when the results of filtering are unsatisfactory.

Amplitude and data discontinuities If high-amplitude spikes are present in the input, a copy of the $f - k$ filter's impulse response will appear within the data itself at each spike location. The educational value of seeing the response in its full glory can seem somewhat trifling when the difficulty of eradicating the contamination is realised. In general, as mentioned in Section

3.4.2, amplitude abnormalities should be removed prior to 2-D processes and amplitude levels balanced. The edges of the data itself constitute amplitude discontinuities and can be a particular problem when a shot gather is $f - k$ domain filtered. If the number of traces within the shot gather is equal to or slightly less than the multiple of two, which is required for fast Fourier transformation, wrap-round can occur, whereby the impulse response spreads data into traces at the opposite spatial extreme of the gather. A particular menace is the very high-amplitude refraction and direct arrival energy on the far traces of the cable which can result in strong contamination of the near traces at the same two-way travel time. Removal of most of this energy prior to filtering using a mute which is somewhat more open than any likely stack or velocity analysis mute, is advisable. Wrap-round can also be reduced, usually at appreciable expense, by padding out the shot gather with blank or random noise traces up to the next highest multiple of two. When a very shallow water-bottom is present, wrap-round can also occur from above time-zero and into the latest part of the traces. For those with a fear of mental topology, possible wrap-rounds are best spotted by drawing a shot gather surrounded by replicas as intimated by Fig. 2.53. 2-D filtering with time domain operators does not suffer from this particular problem as it arises during Fourier transformation of the data.

Irrespective of what can be done to alleviate wrap-round, the extreme traces will always be incompletely processed as in array simulation. Muting of one or two inside traces from a reasonable travel time, i.e. when the fold of stack will not be appreciably reduced, or complete deletion is the common remedy.

Given the above caveats, the technique can be extremely effective. An example of the filtering of shot files to attenuate severe cable tug (noise travelling in both directions along the cable) is illustrated in Fig. 3.68. The appropriate filtering zones of the $f - k$ filter are shown schematically in Fig. 3.67.

An example of the application of an $f - k$ filter after stack is shown in Fig. 3.70 and compares to the unfiltered stack, Fig. 3.69.

3.4.3.3 Multiple attenuation techniques

Methods of attenuating multiples fall into two groups.

3.4.3.3.1 The exploitation of differential NMO

Differential NMO techniques rely on the interval velocity of geological layers increasing with depth into the earth, such that multiple reflections travelling correspondingly more of their path in shallower lower velocity layers have greater moveout with offset than the contemporaneous primaries. Consequently they have a lower NMO velocity. This is most often the case, fortunately, but certainly not always so, as for example when there is a velocity inversion. The effectiveness of these

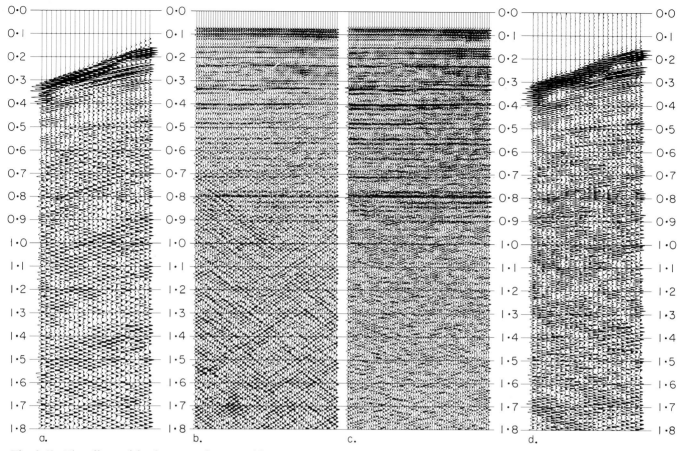

Fig. 3.68. The effects of $f - k$ attenuation on cable-tug on a high-resolution marine line. Parts (a) and (d) show a field record before and after the filtering respectively. Parts (b) and (c) show the data stacked without and with an $f - k$ filter applied before stack. The attenuation zones were 4.5 to 18 msec/trace and -18 to -4.5 msec/trace.

Fig. 3.69. A stacked dataset suffering from a coherent noise problem due to scattering from shallow inhomogeneities.

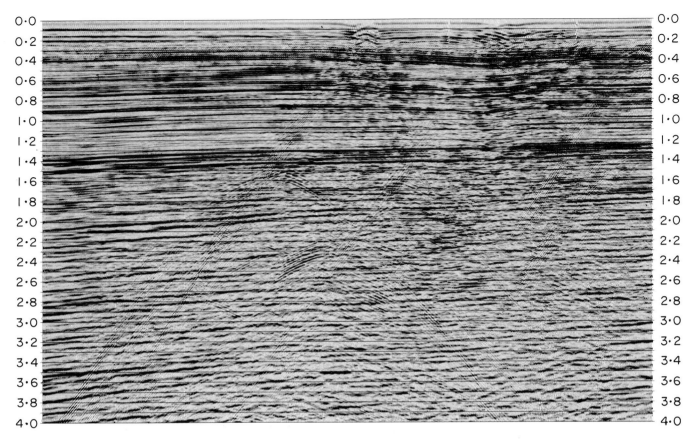

Fig. 3.70. The dataset of Fig. 3.69 after $f - k$ filtering applied post-stack. The attenuation zones were 6 to 25 msec/trace and -25 to -6 msec/trace.

techniques and the stack itself is crucially dependent on the magnitude of the difference between primary and multiple NMO velocities.

Other than the stack itself, there are various methods of achieving exploitation of differential NMO, including:

Weighted stack The discriminatory effect of the conventional mean amplitude stack can be enhanced by weighting the traces appropriately before stacking as a function of their offset. Intuitively, the near traces, on which the multiple generally exhibits little differential moveout, should be downweighted with respect to traces in the middle of the CMP gather. Far offset traces are sometimes similarly downweighted as they are both stretched and often of inferior S/N. Such weights may be calculated empirically, or optimally according to some criterion. The correct stack normalisation following weighting was discussed in Section 3.3.5.3.

Inside trace muting, whereby a time-dependent number of traces are zeroed, can be thought of as an extreme case of the weighted stack. Venturing even further, when multiple attenuation is the overriding consideration and the target zone is deep, a stack of the outside half of the cable is not as outmoded as it may seem.

An example of the effects of a weighted stack is shown in Figs. 3.71 and 3.72, representing a stack without and with the use of weights respectively. Note that both the water-bottom multiple starting at 0.85

seconds on the left and the interbed at about 1.2 seconds in the middle of the section, generated by the primary at about 0.7 seconds, have been effectively attenuated.

Median stacking In a conventional stack, the arithmetic mean of the sample values at the same travel time is used. An alternative measure of the average of a sequence of numbers is the median, which represents the middle amplitude value. The estimator has been termed robust in statistical terms in that it is insensitive to the effects of large isolated numbers or outriders. For example, the mean of 1, 1, 1, 1, 1, 1003 is 168, a rather meaningless number if 1003 was some gross error, whereas the median is 1. Unfortunately, such robust properties have tended to imbue the median with magical properties. In fact, from the point of view of stacking, the median is less robust than the mean in practice. How can this be? The point is that the median is only more robust if each sample is made up of signal and noise. Consider what happens when there is an error in NMO or departures from hyperbolicity due to velocity heterogeneity. At a given reflection time, samples on the far traces will contain no signal whatsoever. Hence, although the mean will always contain some information from the near traces and hence some signal, the median might choose one of the far traces if the NMO errors are sufficiently large, which would result in pure noise.

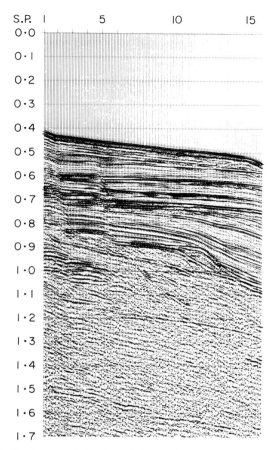

Fig. 3.71. An unweighted stack. Note the water-bottom multiple starting at 0.85 seconds at the left, and the interbed multiple at about 1.2 seconds in the middle of the section. This is generated by the strong primary event at 0.7 seconds.

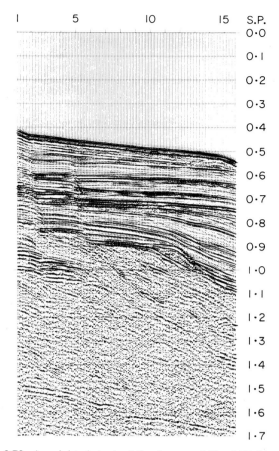

Fig. 3.72. A weighted stack of the dataset of Fig. 3.71. Note the attenuation of the multiples described on that figure.

Having warned the reader, if the data is relatively homogenous in velocity and the velocity is well-known, the median can be an extremely powerful tool.

F − k domain multiple attenuation This technique is particularly popular and exploits the $f - k$ domain separation of primaries and multiples in CMP gathers by selective filtering. The method therefore suffers from similar problems and requires similar consideration as described under $f - k$ noise filtering.

In essence, NMO is applied using a velocity in between the primary and multiple velocity functions. This forces the primaries to have negative dip and multiples to have positive dip. Consequently they lie in different quadrants of the $f - k$ domain. The quadrant corresponding to positive dip is then attenuated with appropriate tapering and the NMO correction removed. The process is exemplified by Fig. 3.73a–d. The effects on the equivalent velocity analyses is quite startling, the region to the low velocity side of the chosen profile being almost completely devoid of energy afterwards. This can be seen by inspecting the velocity analyses corresponding to Fig. 3.73a, d, shown as Fig. 3.74a, b respectively.

Selection of the profile close underneath the primary profile in an effort to attenuate all multiples at lower velocities may not always succeed as expected. Very slow multiples, such as water-bottom reverberation, may remain undercorrected to such an extent that they alias. In this case, a slower intermediate velocity profile will actually aid attenuation, perhaps resulting in two passes of the process being required to attack all multiples. A great temptation when interpreting the velocity spectra is to pick the attractive looking peaks which often appear at velocities very slightly higher than the intermediate profile. Beware! These are generally the remnants of partially attenuated multiples.

A real example demonstrating the power of this method is shown in Fig. 3.75a–b, before and after $f - k$ filtering respectively. The differences are obvious. The difference between the velocity spectra computed before and after are even more obvious and are shown in Fig. 3.76a–b.

A less popular alternative to this technique is one whereby only a single multiple family is attenuated in one pass. This requires the geophysicist to specify the multiple velocity function. Applying the corresponding NMO correction flattens this multiple family such that a small range of dips around the horizontal can be attenuated. After this is done, the data are returned to their uncorrected state. Velocity spectra run after this process show a narrow channel cut through the coherence at the multiple velocity.

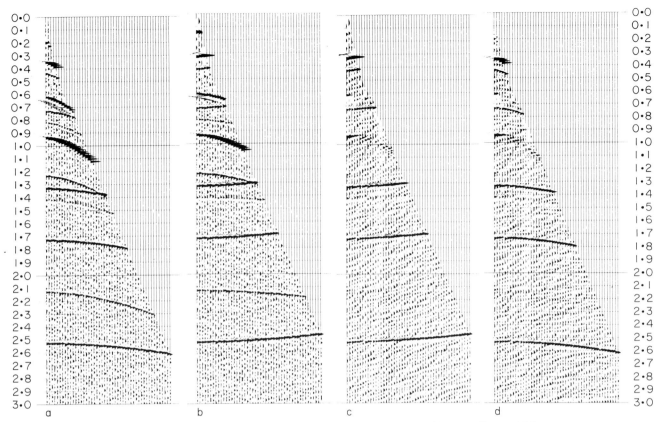

Fig. 3.73. The $f - k$ multiple attenuation sequence of operations applied to a synthetic CMP gather.
(a) The input data.

(b) Application of intermediate NMO.
(c) The $f - k$ filtered version of (b).
(d) The $f - k$ filtered data after NMO removal.

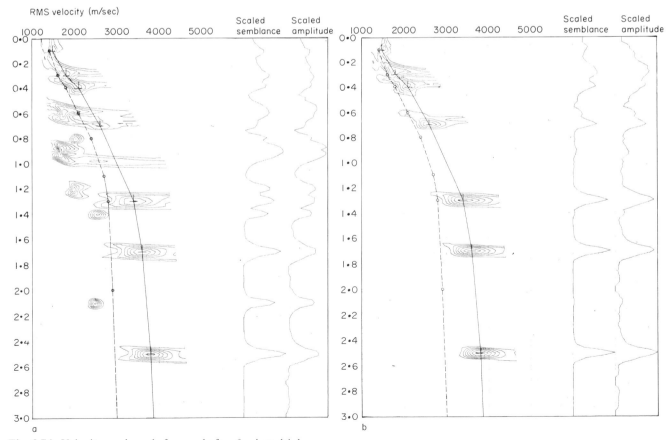

Fig. 3.74. Velocity analyses before and after $f - k$ multiple attenuation.
(a) Velocity analysis done on Fig. 3.73a.
(b) Velocity analysis done on Fig. 3.73d.

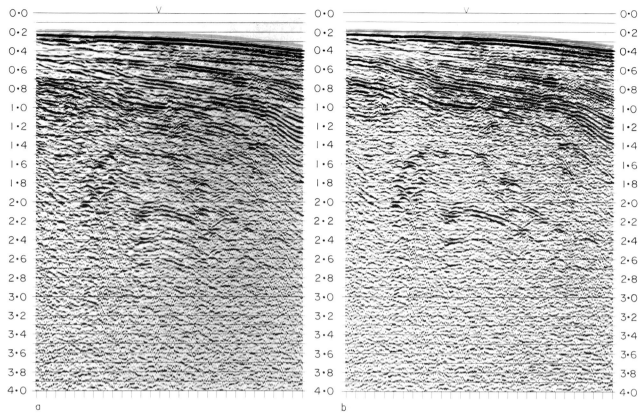

Fig. 3.75. (a) A stack section showing a multiple problem induced by the water-bottom. The tick above the section indicates the velocity analysis position.

(b) The same data as in (a) but stacked after $f - k$ multiple attenuation.

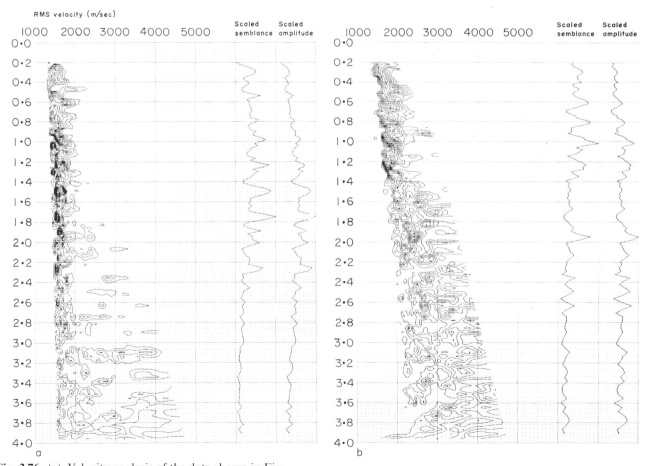

Fig. 3.76. (a) Velocity analysis of the data shown in Fig. 3.75a.

(b) Velocity analysis of the data shown in Fig. 3.75b.

3.4.3.3.2 Exploitation of periodicity

The use of predictive deconvolution to attenuate multiples has already been mentioned and this is almost invariably the method used in deep seismic processing for multiples with periods of 300 msec or less. This often includes water-bottom reverberation. In this case, NMO differential is seldom large enough for the above techniques to be of use and such deconvolution is often adequate. Whenever longer multiple periods occur, still in conjunction with small NMO differential, things get exceedingly tiresome as predictive deconvolution stubbornly refuses to be provoked into working for such periodicities. This is so for several reasons including:

1   The amplitude ratio between generating primary and successive multiples is rarely adequately preserved down the trace pre-stack and not at all post-stack.

2   The multiple period cannot be considered as fixed, especially in the shallow and at large offsets pre-stack. The problem is compounded in regions of complex geology.

Variants of conventional deconvolution have been developed to attempt to overcome these drawbacks. Rather than directly deriving an operator which is convolved with the input trace to produce an output trace, a 'predicted' trace is generated containing multiples only. Before the subtraction of this multiple trace from the input trace to leave a primaries only trace, it is compared to the input trace and adapted in various ways using, for example, correlation. These techniques can be very effective but have achieved some notoriety for the manufacture of phantom events, especially in those areas where high-amplitude primaries dominate the auto-correlation, and close monitoring is recommended.

Less sophisticated techniques also exist for the attenuation of specific multiples which are the geophysicist's equivalent of a pair of scissors. A high amplitude primary, often a deep water-bottom, is digitised and the data is attenuated or zeroed at simple multiples of this period before stack. Many improvements on this basic concept have appeared including the use of some kind of adaptive filtering.

Modern directions in research in this area lie very definitely with studies of the underlying wave-theoretical properties of the development of multiple systems. A considerable amount of work has been done by the Stanford Exploration Project under the direction of Jon Claerbout. These methods show great promise but as yet there remain some theoretical difficulties and immense economic ones.

3.4.4 Cross-section display

3.4.4.1 Introduction

The ultimate test of processing quality lies in the eye of the interpreter. The floating point numbers on the final processed tape need to be converted into a visual display which faithfully reproduces the information contained on the tape in an easily interpretable format. This requires not only good quality plotting hardware, but also an appreciation of the elements of a displayed trace and an understanding of the relation between the parameters specified by the geophysicist.

Display objectives will depend to a certain extent on the interpreter's requirements. For a purely structural regional evaluation, a small scale with emphasis placed on event continuity is acceptable, whereas a detailed stratigraphic interpretation depends on wavelet shape on individual traces being identifiable and clear. Sufficient resolution in display, however, is of general importance and restricts the choice of viable vertical scales. An additional general requirement is that trace to trace overlap of reflection events should result in such events appearing continuous, whilst noise remains non-overlapping, and therefore in the background.

3.4.4.2 Display elements and parameter selection

The majority of quality control (QC) plotting and an increasing amount of final film plotting is currently done in a dot mode as described in Chapter 1. This is illustrated schematically in Fig. 3.77 for a variable area/wiggle display (see below).

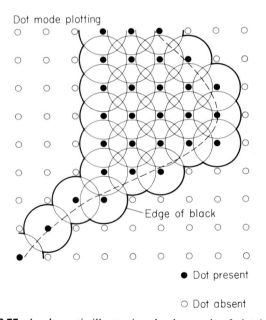

Fig. 3.77. A schematic illustrating the dot mode of plotting.

Electrostatic plotters, so named because they deposit ink at discrete dot positions electrostatically, are often used on-line to display cross-sections as they are produced by the executing seismic job. Stand-alone or off-line plotters which have an internal vector-to-raster converter and a tape drive, are also used to display data directly from a SEG Y tape or similar trace-sequential format. In general, such plotters are not quite so flexible in their plotting capabilities as their on-line counterparts. Electrostatic plotters are by far the commonest kind of plotter in the seismic industry and are currently available with 100, 200, 254 or 400 styli (dots) per inch, the size of each dot being fixed. Although modern electrostatic plotters can also plot directly onto reproducible paper or film as well as electrostatic paper, the invariability of the dot size and quality considerations currently restricts their usage to intermediate QC plots rather than final displays.

Final film plotting is carried out using a more expensive raster graphics device or laser unit. Laser plotters are currently capable of resolution up to 2000 dots per inch although typical seismic applications do not need more than about 500 dots per inch.

Parameter selection is similar for all plotters and consists of the following elements.

Mode Although various methods of displaying what is essentially a matrix of sample values are available, the mode known as variable area/wiggle trace (VAW) is now predominant. In this presentation, positive peaks are plotted solid black and negative troughs as a continuous line or 'white'. The current SEG convention dictates that black positive values should represent a pressure rarefaction in marine data, although the reverse polarity to this is quite common. Other modes available include:
1 Variable area (VA). Peaks are solid black, troughs are left unrepresented.
2 Wiggle only. Peaks and troughs are plotted alike as a wiggly line only (see Fig. 3.79).
3 Variable density. Sample values determine the grey density of regular gridded areas. Black and white televisions, for those of us who remember them, work like this.

Many other combinations or variations of these modes are possible, but are usually reserved for the display of auxiliary attributes.

Scale Certain scales have become standard over the years. For deep seismic surveys, horizontal scales of 1 : 50 000, 1 : 25 000 or 1 : 12 500 are chosen according to survey size and objective, a 'full' scale and 'half' scale often being required. The 1 : 50 000 scale is used for interpretation of regional surveys, whilst 1 : 25 000 and 1 : 12 500 are more likely to be used for the interpretation of potential reservoirs. Most cable group spacings are now metric, resulting in a metric CMP spacing and a whole number of traces per centimetre at the above mentioned scales. For instance, using a 25 metre receiver group spacing, a horizontal scale of 1 : 25 000 results in exactly 20 traces per centimetre. Display programs in seismic software packages usually require the geophysicist to specify the horizontal scale in traces per centimetre (or inch) rather than as a ratio, hence simple but clear calculations are necessary to avoid embarrassing mistakes!

The vast majority of seismic cross-sections produced outside the United States today are at metric scales. Unfortunately, this fact seems to have escaped almost all plotter manufacturers, and most plotters are designed to plot a whole number of traces per inch and are consequently unable to produce an exact metric

scale. This is of course extremely irksome and the only alternative to accepting a slightly inaccurate metric scale is to display at an imperial scale such as 24 or 48 traces per inch and accept a rather more complicated scale of perhaps 1 : 23 662.047. Such is the march of technology. The post-plot survey map should ideally be at the section horizontal scale also.

Accuracy in the vertical scale of a display is of greater importance than the horizontal as the time-tieing of reflections at survey line intersections is performed during interpretation. Ten and 5 centimetres per second of two-way travel time are popular metric scales and 3.75 and 5 inches per second are common imperial scales.

The relationship between vertical and horizontal scales is specified so that a wavelet is easily interpreted. Too large a vertical scale would result in a overstretched appearance whilst too small a scale would make close reflections difficult to resolve. A good rule of thumb is to choose a scale at which a reflection peak width is approximately two trace separations. According to this, reflections of interest with dominant periods between 25 and 50 Hz plot acceptably at horizontal scales of 10 or 20 traces per centimetre and a vertical scale of 10 centimetres per second. Suitable scales for high-resolution data can be calculated from a knowledge of wavelet periods in a similar manner.

Display (gain) level The excursion of a particular sample value from the trace base line can usually be varied about a preset value to enable the amount of trace to trace overlap to be adjusted. The normal criterion used when deciding the gain of a display, is that reflections of interest should overlap to form continuous black bands while noise does not. This clarifies the section considerably. A typical preset or 0 db gain level would be to set the calculated average amplitude of representative traces to an excursion of one-half of the trace spacing. A gain level of + 6 db would then result in a sample value of average amplitude having an excursion of one trace spacing. Irrespective of the geology, gain level is usually chosen within a + or − 3 db range about the preset. A slightly different view of this parameter choice is to consider that it should be made with reference to the background noise rather than the strength of any particular reflections. A level which just fails to result in overlap of the background noise is then the objective, all reflections of greater amplitude than the noise exhibiting some overlap and therefore lateral continuity. See Fig. 3.78.

Bias The position of each trace boundary determines a base line which can be used to plot zero amplitude. All positive trace amplitudes are then displayed black and all negative ones as 'white'. Plotting in this manner, however, gives little visual prominence to reflections which consist of a dominant negative lobe. To equalise the contrast or 'standout' between significant positive and negative reflections, a bias is specified which moves the zero amplitude level away from the trace base line.

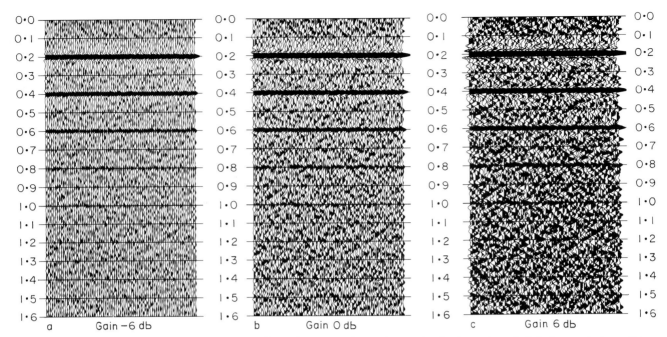

Fig. 3.78. Sections showing equally spaced events throughout the time section. Amplitudes are reduced by 3 db per event. Display (b) is a standard gain setting. Displays (a) and (c) shows the result of gain settings 6 db below and above this level.

In the computer, this is achieved by adding a desired constant amplitude to each sample value before display. The addition of bias thus increases the average grey appearance of the section. Specification of bias is normally done as a percentage of trace spacing, values in the range 0–35 per cent being common.

Clip level Very high amplitude reflections are generally clipped at a specified excursion of between 2 and 4 times the trace spacing to prevent nearby adjacent traces being obscured or confused. Some clipping may also be imposed for reasons of economy in program design.

Interpolation method On both analogue and digital (dot-mode) plotters, the trace needs to be interpolated between data points. For example, for a plotter resolution of 100 dots per centimetre, and a vertical scale of 10 cm per second, traces with a sample interval of 4 msec will correspond to a sample spacing of

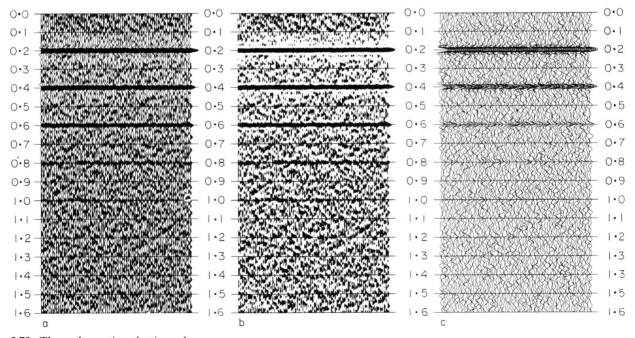

Fig. 3.79. Three alternative plotting schemes.
(a) Variable area wiggle with 30 per cent trace bias.
(b) Variable area only with 15 per cent trace bias.
(c) Wiggle only.

0.4 mm, but stylus positions on the plotter correspond to a spacing of 0.1 mm.

The quality of some interpolation schemes has already been discussed in Section 2.9.6, and in cases where dominant wavelet frequency is in excess of half the Nyquist, the greater fidelity of sinc function interpolation or its equivalent may well be necessary.

Dot spacing and size The display elements discussed so far have been common to both analogue and digital plotters. The choice of dot spacing and size is, of course, peculiar to digital plotters and has considerable influence on the resolution and general appearance of a section display.

An obvious requirement is that the relation between size and spacing results in a continuous image. If the dot size is chosen excessively large, however, degradation of resolution occurs. A dot diameter somewhere in the range of 1.5–2 times the dot spacing is satisfactory. Dot spacing is, to a large extent, determined by economics, but a dot density greater than 120 dots per cm is sufficient to perceive continuity. For small scale displays or high trace densities, a smaller dot and commensurate spacing are required to retain resolution.

3.4.4.3 Trace balancing prior to display

Average trace amplitude will, in general, vary considerably along a line according to geology and local noise conditions. Temporal variation is also likely even after compensation for any residual decay using a fixed scaling function. These variations are often such that at any given display gain setting, certain data zones are either of too high or too low amplitude for successful interpretation. Temporally and spatially variant data adaptive balancing is required before display.

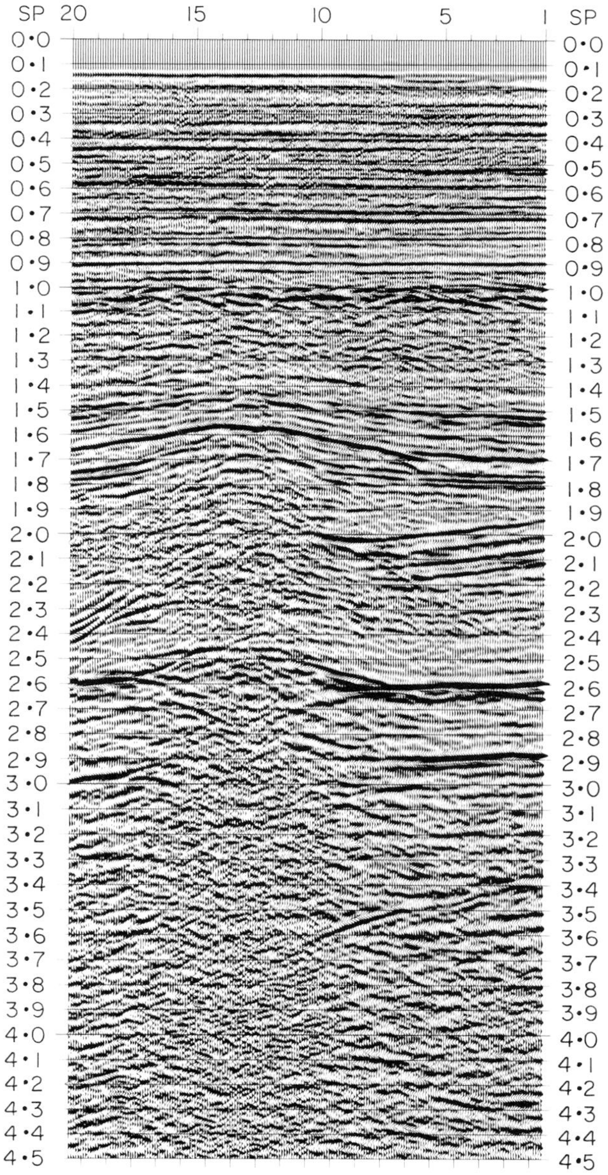

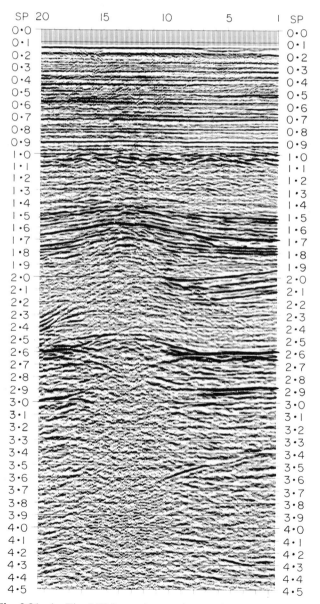

Fig. 3.80. A standard marine section with a cosmetic AGC for display purposes. Note the blank zones on either side of strong reflectors such as that at around 2.6 seconds on the right of the section.

Fig. 3.81. As Fig. 3.80 but using a robust estimator for gain computation.

When using data dependent gain routines as described in Section 3.3, great care must be observed in order not to over-equalise by use of too short a sliding window. *S/N* will always be degraded using an adaptive gain as zones of high amplitude (reflector sequences for example) will be scaled down to a similar amplitude to zones predominantly containing noise.

The problem is particularly serious when very high amplitude reflections such as some unconformities or deep water-bottoms are located close to lesser amplitude reflections of interest. The average amplitude of the zone about the major reflector will be severely reduced due to its effect on the calculation of the gain scalar within the window. Such effects can be ameliorated by using more robust estimators of window average amplitude. An example is shown in Fig. 3.80 whereby minor events around the strong reflector at 2.6 seconds between shot points 1 and 10 are significantly affected. The same calculation using a more robust estimator is shown in Fig. 3.81.

3.5 Marine seismic data processing schemes

In this section some of the many issues involved in the practical processing of seismic data will be described.

The techniques described in the preceding section have various purposes all aimed at enhancing the geological interpretability of the final display section. The subset applied to any particular line is determined through a combination of theoretical knowledge, experience and parameter testing. However, for a given survey, the sequence of processes and key parameter selection will remain constant from line to line to enable data comparison and reflection tieing. A variation in wavelet shape or multiple content for example, is certainly to be avoided.

The processing sequence of Fig. 3.82 illustrates all necessary processes, along with some of the more common optional ones, which are applied to multiplexed field data. The mandatory processes shown in solid boxes are in a fixed sequence. Most optional processes can vary in position within this fixed sequence to an extent, but simple processing sequences employed by different companies are usually similar. This probably reflects an intuitive understanding of the points discussed in Section 4.5 concerning commutativity.

3.5.1 Process ordering

The gather and stack processors are important in determining the position of optional processing steps, particularly 2-D processes, as a change in trace ordering occurs. Coherent noise rejection filtering and array simulation for instance, can be used to attenuate noise trains which form linear patterns on common shot files and hence need to be applied prior to CMP gathering. Multiple attenuation techniques which exploit differential velocity moveout are normally performed on gath-

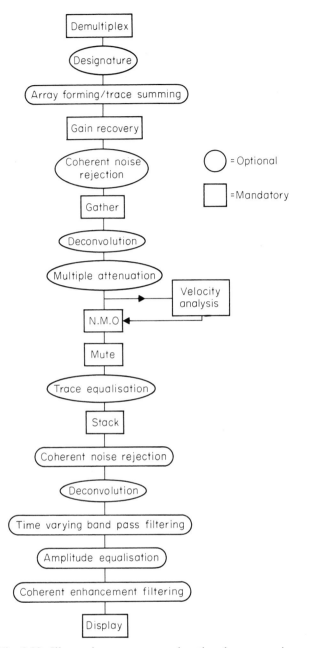

Fig. 3.82. Illustrating necessary and optional processes in a typical marine seismic data processing sequence.

ered data. However, single channel processes such as deconvolution, scaling and muting, can be inserted into the sequence with more freedom although often the relative position of these steps is vital.

3.5.1.1 *Designature or signature deconvolution*

This is normally the first step following demultiplex, but must always precede any explicit amplitude correction.

3.5.1.2 *Array simulation*

As this stage effectively simulates a different field configuration, it should appear very early in the sequence. Its position, relative to designature, is generally unimportant unless some kind of non-linear technique, for example, time-variant moveout correction, is used. Bad

To summarise, such shallow targets present fundamental problems which require much attention in order to make the structure visible. A side effect is that the niceties of wavelet shaping are often rendered irrelevant.

Sand/shale sequences Many regions comprising sand/shale sequences provide data suitable for some of the most exciting stratigraphic processing techniques for the detection of hydrocarbons, based on trace attributes such as amplitude envelope values and instantaneous phase and frequency as defined in Section 2.6.7. Most features of the data vary gradually in both time and space. The need to discover stratigraphic traps places great emphasis on wavelet control and knowledge and attenuation of short period multiples.

Salt diapirs The diapiric upwelling of salt into dome-like structures with very steep sides is frequently associated with major distortions in the surrounding sedimentary pile. Also, salt is a high velocity medium whereas surrounding sediments usually have much slower velocities. These factors lead to many problems and include the following:

1   Velocity interpretation can be very complex. Stacking velocity rises rapidly as dip increases on the flanks of the structure. (See Section 3.3.6.1.) When salt structures are short in lateral extent compared to the cable, ray paths become contorted and the hyperbolic moveout and CMP assumption break down. Consequently, conventional stacking of reflectors below the base of the salt is particularly disappointing as they are often well resolved away from the deforming effects of the salt.

If conventionally stacked data is produced, the geophysicist should expect to pick highly variable velocity functions in order to stack reflections optimally below the salt. In particular, functions below the flanks of domes will possess what may seem a totally unreasonable shape due to ray-path distortion. Underlying reflections will also experience considerable 'pull-up'; that is they will appear much earlier in time than their extension away from the salt, owing to the latter's high velocity compared to the sediments it has displaced.

Velocity analyses are obviously required at short intervals or careful spatial positioning performed. In particular, analysis positions should be selected on the flanks of the dome as well as the highs and lows of structures. When target zones are beneath the salt, horizon-orientated velocity analyses should be considered if economics permit. See also Section 4.4.

2   The breakdown of the CMP assumption leads to attempts to overcome the problem by using either the 'special processing' techniques of layer-replacement, whereby the usage of appropriate time shifts eliminates the effects of the variable velocity overburden, or some kind of migration before stack.

3   Salt deformation often results in faulting over the top of the structure and shallow diffraction features may proliferate. The hyperbolic tails of such diffractions are often a serious problem in the target zone. Post-stack

dip filtering may not always be possible without attenuating the primary reflections above the base of the salt. If the target zone is below the salt and dip separation is possible here, a temporal merging of filtered and unfiltered data could provide a solution.

4   The presence of high-dip, high-impedance contrast interfaces can produce severe multiples in a deeper target zone which, because of their dip, have a stacking velocity very close to the primary stacking velocity. Differential NMO-based multiple attenuation techniques are therefore inapplicable and periodicity cannot be relied on for long-period multiples generated by such interfaces. A suitable nicely-worded apology to the interpreter may be the only course left open.

5   The velocity variation can cause immense problems in migration. As will be described in Section 4.3.1, careful velocity modelling is required in order not to produce rubbish. The fact that the stacking velocity distribution around a diapir bears little resemblance to anything physical certainly does not help.

3.5.5 Quality control procedures

Displays can be obtained after each process if required, to check that everything has performed adequately. In general, this is uneconomic in practice and displays are produced only at strategic points in the processing sequence as the response of the data to many processes is uniform and predictable. This is particularly true when dealing with 'production' lines, that is the remaining lines of the survey which are to be processed in a similar manner to, and using the key parameters established from, the test line or lines. Attention is paid to line dependent features such as possible noisy acquisition channels and the correctness of NMO removal, as well as a general check being made to confirm the effectiveness of the test line parameters on each production line.

A rudimentary quality control scheme would employ displays as follows:

Stage 1

Near trace displays (cf. Fig. 3.28) Near traces from all shots along the line after data-independent amplitude decay correction are displayed to give an early indication of the gross geological structure and water depth trend. On the assumption that this trace is representative of the complete shot gather, the geophysicist may:

1   Delete the complete shot if a misfire is suspected, or noise conditions are excessive.
2   Make timing corrections if the relation between the shot and the time of the first scan recorded is abnormal.
3   Confirm the offset of the cable from the shot by measuring the direct arrival time.
4   Check that there was no obvious unadvertised change of acquisition parameters, such as filter setting, along the line.

This display may also be used if data quality is sufficiently high to determine optimum locations for velocity analyses. Used in conjunction with demultiplex diagnostic printout and the observer's logs, it can also be used to identify and correct deviations from standard shot spacing.

Selected shot gather displays (cf. Fig. 3.36) Displays of all traces of regularly spaced shot gathers after data-independent amplitude decay correction can be used to indicate:
1 Consistently bad or reversed polarity channels.
2 Some coherent noise problems and range of dips involved.
3 Approximate interval velocities of shallow layers using refracted energy.
4 Potential multiple problems.
5 Interference over narrow frequency bands.

Stage 2

NMO corrected CMP gather displays (cf. Fig. 3.50) These are extracted at, and possibly between, velocity analysis positions, preferably before muting but with the mute pattern shown. Their usages include:
1 A check on the accuracy of NMO correction. Primary events should align horizontally. Special attention is paid to gathers between velocity analysis positions in geologically complex areas as linear interpolation of the stacking velocity field between positions may not be suitable. Additional specification between original analyses may be required according to structure. Alternatively, other methods of interpolation may be considered, such as iso-velocity, which effectively interpolates along constant velocity contours.
2 Offset mute checking. The reasonableness of the selected mute pattern along the line is monitored. Changes in geology and related interval velocity can result in unwanted refracted energy being left in, or excessive NMO stretch. This will require adjustment to the pattern.
3 Amplitude monitoring. Irregularities in trace amplitudes which would lead to a less than optimal stack being investigated.

Raw stack display (cf. Fig. 3.40) A raw stack display is often the earliest opportunity to observe the geological structure along the line with any certainty, and as such may indicate that previous parameter selection such as deconvolution window timing or velocity analysis positioning was unsatisfactory. For this reason in poor data quality areas, a brute stack using a representative velocity profile is produced at an early stage. As has already been mentioned, the brute stack can be a much better alternative to a near trace display for aiding the interpretation of velocity analyses.

A display of the raw stack should always be scanned for any irregularities in geological structure which may be attributable to processing problems. Disappearing horizons or horizons with missing segments are particularly difficult to get past the vigilant interpreter even when assisted by extravagant free lunches. Possible

causes include the attenuation of primary energy during coherent noise or multiple attenuation over localised areas, incorrect stacking velocity interpretation, or possibly, the use of too open a mute.

A change in frequency content (i.e. reflection character) which appears at a consistent two-way time unrelated to geology or amplitude level, invariably means that a merge zone between deconvolution or other operator application windows has been of insufficient length. In hopefully rarer circumstances, this may be due to inadvertent premature termination of a pre-stack process or even mis-specification of a gain curve, in other words, a simple job set-up error.

This basic scheme will, of course, be much enlarged for test lines or detailed processing with extra displays obtained both prior and following many processors. When shot by shot designature with recorded signatures is necessary, a monitor of every signature should be standard practice.

3.5.6 Testing schemes

Prior knowledge of the region and inspection of field files is likely to determine which processes require testing pre-stack. Some processes and parameters however, are routinely tested over a short data panel.

3.5.6.1 DAS/DBS

Decisions regarding deconvolution before and after stack and space- and time-variant filtering, are obviously closely related and dependent on other spectral modifying processes such as designature. Testing and selection of the individual parameters cannot, therefore, be performed in complete isolation. This would suggest an approach whereby each stacked DBS test panel is forwarded as an input to the full DAS test suite, each panel of which is used in turn as input to a filter test suite. All possible combinations of parameters which affect spectral content can then be assessed. A quick calculation of the number of panels thus required (of the order of $6 \times 6 \times 6 = 216$) soon convinces the aspiring processing geophysicist and hopefully the interpreter, that this is not practicable. A reasonably chosen subset of panels is one option available, for example, spiking deconvolution both before and after stack is unlikely to be worth testing. If a broad band filter only is applied to all the DAS panels produced, and the DBS/DAS combinations suitably restricted on theoretical and common sense grounds, a suite of less than 18 panels results. The choice of DBS and DAS is made on this basis and a single filter panel suite obtained.

Even this restricted DBS/DAS suite may be thought of as haphazard and rather short on geophysical input. When faced with a large number of test panels derived from combinations of different parameters, a reasoned geophysical justification for any particular choice is often hard to make. A step-by-step approach to parameter selection, whereby the choice of DBS is made prior to any DAS trials, does allow a larger range of parameters to be tried and may prove easier to analyse. If, at this stage, two DBS trials are deemed to be equally promising and theoretically valid, then these panels can be forwarded to a restricted DAS suite. These approaches can be summarised as follows:

1 Initially determine all the DBS and DAS combinations which are thought plausible, and choose a broad band filter which reduces excessive amplification of very high and low frequencies with poor S/N ratio caused by a short gap deconvolution operator. Select DBS and DAS parameters simultaneously.

2 Test a full suite of DBS options and select one, or possibly two, panels to use as input to a full DAS suite, a broad band filter being applied to all data prior to display. Then make the choice of DAS.

In the past, DBS decisions have been made on the basis of unstacked data for reasons of economy, but nowadays, it is more usual to stack CMP gathers in 2 to 5 km panels across representative geology. A 'standard' suite of trial parameters for deep marine seismic data sampled at 4 msec would look as follows:

Panel	minimum a.c. lag (msec)	maximum a.c. lag (msec)	a.c. design windows
1	4	w.b. + 60	1
2	12	w.b. + 60	1
3	20	w.b. + 60	1
4	32	w.b. + 60	1
5	w.b. − 60	w.b. + 60	1
6	20	100	1
7	20	w.b. + 60	2

Where w.b. is short for water bottom and a.c. for auto-correlation. Note the differences between this and a strategy for high-resolution marine data as exemplified by Fig. 3.61.

Of the above: panels 1 to 4 test different degrees of wavelet compression whilst also performing attenuation of water-bottom multiples; panel 5 tests water-bottom attenuation only; panel 6 tests wavelet compression only; and panel 7 tests for the effects of multiple a/c design windows. A key feature of this example suite is that all panels deviate in respect of only one parameter from a best guess panel (in this case number 3). Although the best guess panel may be substantially different from that in the example above, it is crucially important to be able to relate any changes in the data to a single parameter.

DAS test suites commonly look roughly similar to DBS suites. If wavelet compression only is required, a shorter maximum a/c lag may be used when partial compression has been achieved before stack. Some parameters may not be tested according to choice of DBS.

Example DBS and DAS suites are shown in Figs. 3.61 and 3.87 respectively. In this case, as the high-resolution 'shallow' dataset is being used, the parameter values differ from the 'deep' seismic case given earlier. The auto-correlations from each trace are also displayed to aid analysis of inherent periodicity in the data in light of the comments made in Section 2.7. The first 300 msec of the auto-correlation lags are shown at an enlarged scale to facilitate the choice of the correct periodicity.

The panels in each suite show a 1–100, 2–100, 4–100, 8–100, 8–40 and 40–100 msec minimum and maximum prediction distance respectively. The same amount of white light was used in both suites.

One other point worthy of mention is the difference in the amplitude of the noise background on the 1–100 msec filter (left-most panel) between DAS and DBS. On the DAS, the noise is much larger in amplitude. This relates to a point made earlier to the effect that because the S/N is inherently worse before stack, the data itself contains sufficient random noise to control gain of the deconvolution filter. This is not true after stack, the white light added being insufficient on its own to accomplish this for a 1 sample lag or spiking deconvolution.

3.5.6.2 Offset muting

Mute patterns can often be adequately defined from inspection of large scale CMP sections. For more rigorous testing, stack panels can be obtained containing progressively more traces, and the number of traces live at set times at which data quality is optimum, noted. For 48 fold data, trial fold of stacks of 1, 2, 4, 6, 8, 12, 16, 24, 36 and 48 would be sufficient.

An alternative is to define a sensible mute pattern and calculate patterns whereby fold of stack at any time deviates from this by a fixed percentage. Stack panels are again produced using the various mutes.

3.5.6.3 Time-variant filtering (TVF)

A series of band-pass filters defined by the geophysicist are applied to a stack panel following DAS to produce a suite of panels which aid in the selection of optimum time-variant filters. There are three common types of test:

(a) *Broad band filters similar in cut-off and slope to those which will eventually be applied to the data (see Fig. 3.64).* Panels which range from low-frequency band filters suitable for use deep in the section to high-frequency band filters appropriate to shallow data are plotted, and parameter choice is simply a matter of selecting the best looking panel at any time. The disadvantage of this kind of display is that degradation of data quality can be either due to an unsuitable high or low cut frequency. Similarly, this approach does not easily indicate the filter bandwidth which should be used.

(b) *Narrow band filters with relative amplitude preserved (see Fig. 3.63).* Using a series of contiguous or overlapping narrow band filters (say 1/2 octave), a visual impression of the spectral distribution of both signal and noise can be obtained. Time-variant low and high cut frequencies are picked independently. As no display is obtained similar to the final section in this

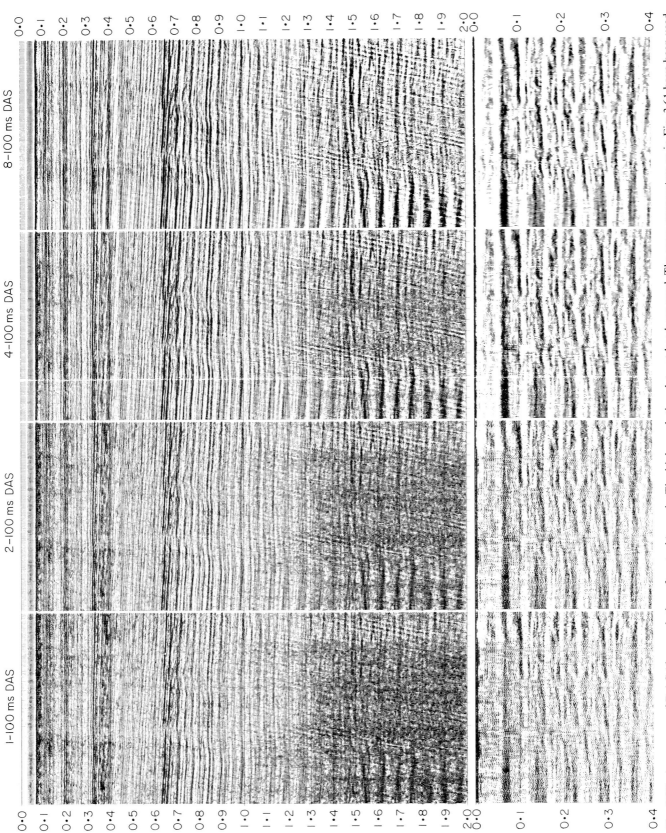

Fig. 3.87. A sequence of DAS (deconvolution after stack) panels. The high-resolution dataset has been used. The same parameters as in Fig. 3.61 have been used. Auto-correlation traces are displayed beneath the data. Panels continue below.

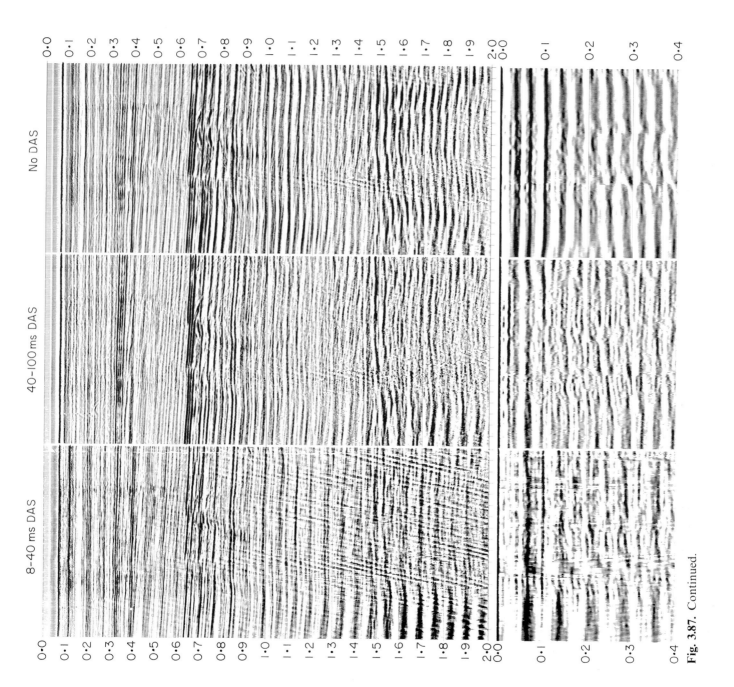

Fig. 3.87. Continued.

method, broad band filter panels as in (a) above are often computed in addition.

(c) *Variable low/high cut filters with fixed high or low cut respectively.* Panels are produced in much the same way as (a) except that one end of the filter spectrum is fixed whilst the other is varied. An increment of 5 Hz for the high cut and 2 Hz for the low cut is reasonable. The time-variant cut-offs are again produced independeltly. The fixed frequency cut-off should be well outside the signal band.

As a practical comment, a filter band of 24–140 Hz in Fig. 3.63 effects a good compromise, removing both the low-frequency cable noise and passing the highest frequencies with signal content.

3.5.6.4 Final amplitude balance and display

There is no recognised test structure for final gaining and display but the following will probably require some trials to be run:

1 Data-dependent scaling (for example, window length).
2 Display gain level (db setting).
3 Polarity.

3.5.6.5 The migration

Migration is discussed in considerable detail in Chapter 4. Its short appearance here is merely to complete the sequence of the processing of the high-resolution dataset. A finite-difference time-migration of the filtered stack of Fig. 3.85 is shown in Fig. 3.88. The velocity is relatively laterally homogeneous in this example, so no dramatic effects are to be expected. However, note the clarification of the faults.

3.5.7 Section annotation and labelling

3.5.7.1 General

Following the processing and display of seismic data, the important final step is that of annotation and labelling. Currently, it is quite common for the final display to be created together with top header annotation. In addition, a side label may also be added automatically, using information supplied with the present and previous computer runs in the sequence.

The side label itself is usually of A4 width or less, with a depth up to the length of the data. Increasing the overall length beyond this is not considered desirable.

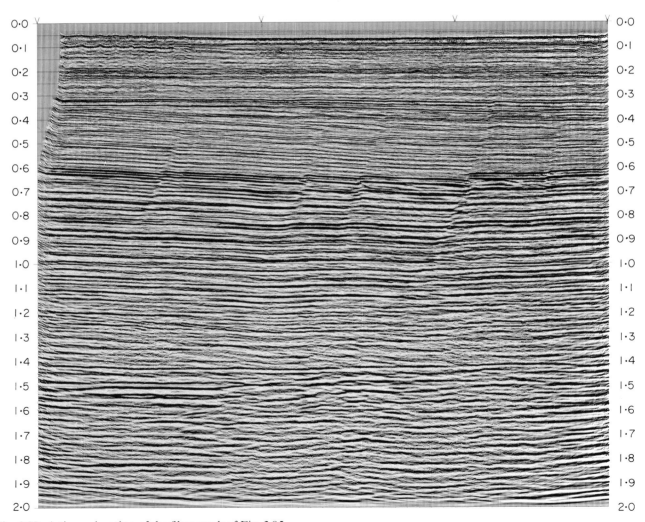

Fig. 3.88. A time migration of the filter stack of Fig. 3.85 using the finite-difference method. Note the clarification of the faults.

General identifying information is placed in the top half of the label allowing the paper section to be fan-folded to A4 size for archiving. Such information includes the line number and shotpoint range, processing title, survey area, and names of the oil company and acquisition and processing contractors as applicable. An abbreviated label is also situated at the opposite side of the section, usually parallel to the edge of the display in order to give instant identification when filed in cabinets.

3.5.7.2 Location information

Knowledge of the position of the line both absolutely and relative to other lines in the survey, is of great importance and appears in the labelling in various ways:

Generalised location map. This shows the location of the survey relative to coastlines and lines of latitude and longitude.

Survey map. This shows the orientation of lines to each other. The line in question is highlighted.

Direction indicator. This indicates the direction of shooting of the displayed line, usually by a compass heading and arrow.

Shotpoint dependent labelling. Annotations are made along the length of the line, usually directly above the top of the traces, defining shotpoint positions. As a general point, all efforts should be made to make this annotation entirely consistent with the shotpoint location map (sometimes known as post-plot), both in

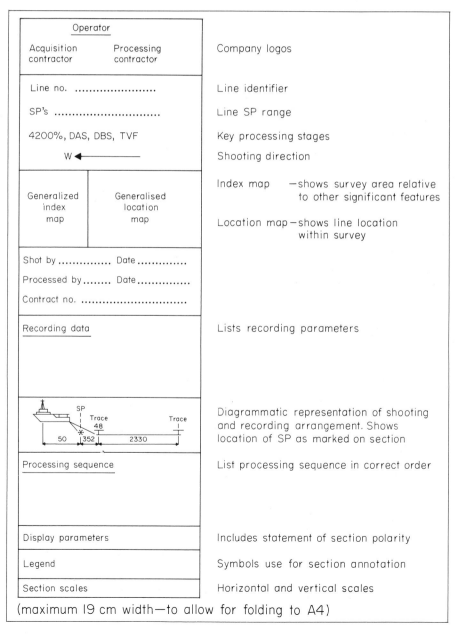

Fig. 3.89. Typical side label format for section displays.

precise line labelling and shotpoint labelling and ticking. If parts of the same line have been reshot, shotpoint numbering should not be duplicated. Errors of this nature during acquisition should be corrected, usually by the addition of 1000 or 10 000 to the shotpoint number, both on the seismic section and shotpoint map. Other shotpoint related variables such as water depths, may optionally be displayed at higher levels above the display. CMP and other ensemble numbers are usually relegated to the bottom of the section, their importance being to the processing stage rather than the interpretation.

The highest two levels of annotation comprise intersection and velocity information. An additional intersection tick at the bottom of the sections aids accurate folding when line tieing. Velocity information usually consists of stacking velocities (commonly and mistakenly called RMS velocities) and interval velocities between picked horizons, calculated using the Dix equation and assuming that stacking velocities and RMS velocities are equivalent. On migrated data, the migration velocities should be displayed. On much seismic data these, too, are often the stacking velocity. This is discussed further in Chapter 4. Finally, depths also appear, as calculated from the interval velocities. These should be taken with several large pinches of salt!

3.5.7.3 Acquisition details

Full details such as source type, depth and energy, filter settings and geometrical distribution and the like are set out in an information block together with a diagram illustrating the cable configuration and distances. The primary source of this information — the observer's field log — has an unfortunate habit of getting misplaced over the years and reprocessing efforts sometimes rely solely on the side label for acquisition details. Fortunately, more and more parameters are now being recorded into field tape headers.

3.5.7.4 Processing details

Following the acquisition information block, the processing stream is described. Key parameters are set down in the hope that the final section could be recreated if required at some future time. Unfortunately, the complexity of modern processing sequences coupled with the habit of using proprietary and totally obscure process names, means that only an approximate reconstruction can realistically be achieved. In practice, the greatest advantage of recording processing details is that features of the display attributable to processing techniques and decisions can be noted and treated accordingly.

Guidelines governing layout and information content are occasionally issued by august industry bodies such as the UK Offshore Operator's Association (UKOOA). A recommended side label format issued in 1980 by the Norwegian Industry Association for operating companies is shown as an example in Fig. 3.89.

3.6 Irregular and extended geometry

3.6.1 Introduction

One of the most distinguishing features of the simple 2-D marine processing considered so far, is the regularity of geometry as was described in Section 3.3.2. This is not the case for other kinds of processing such as 2-D land processing and 3-D land or marine processing. Consequently, for these latter kinds of processing, geometrical considerations consume a much greater proportion of both the geophysicist's and the computer system's efforts. In addition, errors in geometry specification tend to be harder to spot and have a considerably more deleterious effect on the final product. It is worthwhile therefore, to spend a little time on this oftneglected and unglamorous, but vitally important part of processing.

As a prelude, it is useful to define the following categories of dimensionality to cover common concepts in seismic data processing:

1 1/2-D seismics Bore-hole seismic acquisition and processing.

2-D seismics Standard reflection seismic acquisition and processing. No account is taken of cross-line effects in either.

2 1/4-D seismics Acquisition is conducted along a single non-straight (crooked) line. 3-D techniques are used in processing to make simple estimates of crossline parameters such as the cross-dip of particular horizons.

2 1/2-D seismics Commonly called 3-D seismics. Acquisition consists of shooting a number of closely adjacent lines as 2-D lines, and 3-D processing techniques are used to merge all the data into a 3-D volume in the computer. To pander to current tastes in jargon, this category will be referred to as 3-D seismics here.

3-D seismics Acquisition is also 3-D, involving an areal spread of receivers for each shot. Processing is 3-D. Such surveys are still rarely carried out because of expense and will not be considered here.

All of the issues discussed in Section 3.3.2 are, of course, relevant. As will be seen, irregularity implies that the concept of the common mid-point has to be dropped. It might be argued that 3-D marine data by virtue of its essentially 2-D acquisition is regular, but this is definitely not the case as will be seen later.

Historically, irregular 2-D geometry was developed in parallel with regular 2-D geometry. Although regular geometry is, of course, a subset of irregular geometry, in practice, they are treated differently and normally by different computer programs for reasons of efficiency. 3-D acquisition and processing evolved later and will be considered separately, although the difference is essentially one of scale rather than of methodology.

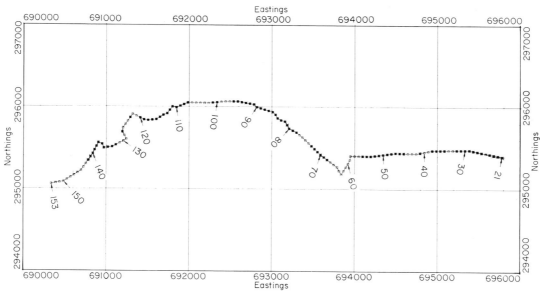

Fig. 3.90. A plan view of a typical crooked land line. Shot positions are indicated by dark squares and receiver positions by light squares.

Before getting into the details, the following dictum is well worth absorbing: all else being equal, the more regular the acquisition geometry, the better the final processed data.

3.6.2 The crooked line and binning

The crooked line concept has to be carefully defined as it has both a trivial meaning and a non-trivial meaning. In essence, a crooked line is one which is not straight! This is the trivial meaning. In practice, 'crooked line processing' has come to mean processing which exploits the crookedness to gain geological insight in the cross-line direction. This is the non-trivial meaning. In the nomenclature introduced in Section 3.6.1, it corre-

sponds to 2 1/4-D processing. Of course, irregularity may either be inevitable or deliberately used to provide such cross-line information, however, irregularity is generally inevitable in 2-D and 3-D land acquisition due to a mixture of terrain and access problems.

By way of illustration, Fig. 3.90 shows a fairly typical plan view of a land line. The dark squares indicate shot positions and the open squares indicate receiver positions. What effects does this have on the data? The most important issue was hinted at above and is the demise of the common mid-point assumption. To see how hopeless such an assumption is on this line, Fig. 3.91 shows the same plan view as Fig. 3.90 with the mid-points between each source–receiver pair superimposed. As can be seen, between eastings 691500 and 694500, mid-points are badly scattered.

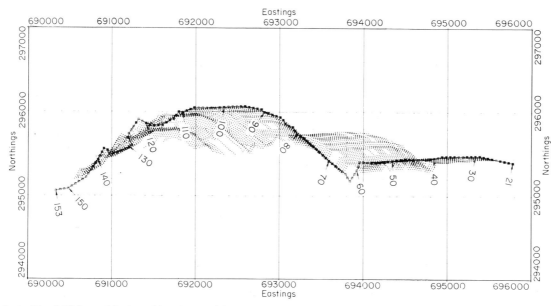

Fig. 3.91. As in Fig. 3.90 but with the mid-point positions between the sources and receivers indicated.

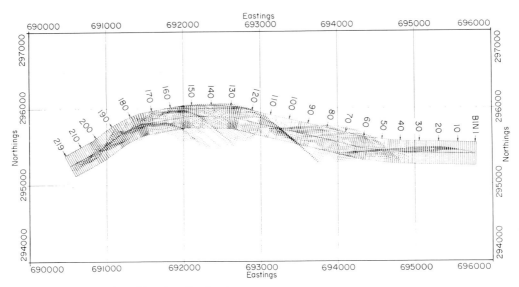

Fig. 3.92. A 'binning strategy' for the mid-point distribution of Fig. 3.91.

The universal solution to this scattering is to extend the definition of the common mid-point to cover a small area. The area has come to be known as a 'bin', a term used in other physical sciences such as radio-astronomy with the broadly similar meaning of a logical container in which items of a 'similar' nature may be grouped. Bins come in a variety of shapes but rectangular ones are by far the most common. The size, and also aspect ratio in the case of rectangular or elliptical bins, have an important role to play. The skill is to choose a bin size and shape at a point on a line which gives a sufficiently large bin population of suitable traces, i.e. number of contained source–receiver mid-points, without being large enough to reduce the spatial resolution. Suitability will be considered shortly. A 'binning strategy' for the line of Fig. 3.90 is shown in Fig. 3.92, and an enlarged section of this in Fig. 3.93. Note that this particular strategy simply chooses evenly sized bins with no overlap.

The alert reader will have noticed that the diagrams show surface mid-point positions rather than sub-surface positions as were shown in the simple introduction to geometry given in Section 3.3.2. Land geometry is usually sufficiently complicated for this to be the norm. In addition, other surface effects such as time delays due to topography, have a major influence on land processing, and the land processor must simultaneously correlate surface information of various kinds in anticipation of likely problems.

Note also that Fig. 3.92 shows bins chosen to lie on a line which approximately best fits the mid-point distribution. This line is, of course, the processing line and is the surface position to which the final section will relate. The processing line may actually be anywhere, but common sense usually dictates that it behaves as described above unless there are good reasons to choose otherwise.

The displays shown in Figs. 3.90–3.93 are not sufficient to allow processing to proceed. A most important

addition is to perform some kind of statistical analysis on the contents of the various bins to assess 'suitability' of their contents. Before enlarging on this, it is worthwhile trying to anticipate some of the effects of irregularity and evaluate the criteria for suitability.

Bin centre, trace centroid and dislocation Referring back to Fig. 3.93, the binning strategy shown, results in bins whose geometric centres are evenly spaced along the processing line. On the other hand, it can be argued

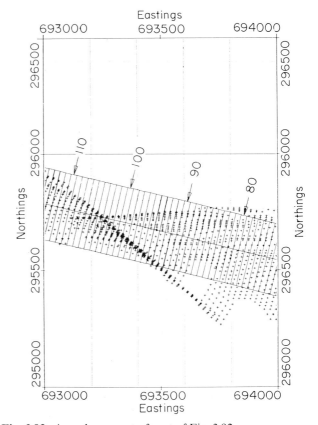

Fig. 3.93. An enlargement of part of Fig. 3.92.

that the trace centroids or mean positions of the traces lying within the bin should be taken as the bin centres. The distance between the trace centroid and the bin geometric centre is sometimes known as the bin dislocation. This issue is both controversial and important as the stacked trace obtained by stacking all traces within a bin is deemed to lie at the bin centre. In either case, some kind of interpolation is involved, either explicitly or implicitly, as the final section must be regularly spaced both for visual display and most kinds of migration. It is certainly desirable to keep dislocation to a minimum.

Trace offset distribution It is most important for a number of reasons, notably velocity analysis and multiple attenuation, to have an even spread of offsets within a bin. It is obvious that a bin containing only small or large offsets will be of very limited use. It is also equally important to have a reasonably uniform fold or bin population as otherwise such amplitude sensitive processes as migration and residual statics might be detrimentally affected.

Trace azimuth distribution This is a little more subtle. In cases where there is modest strike, the angle of the source–receiver pair defining a particular mid-point with respect to some reference direction is not particularly important. However, when strike is significant, traces of the same offset but different azimuth will have different two-way travel times to the same horizon. In practice, this is a rather expensive effect to accommodate as it usually requires a fairly thorough first-pass through the data to define some kind of strike or 3-D model in order to incorporate the effects on a second iteration. It is therefore a good idea to restrict the distribution of azimuths as much as possible. This is exactly the opposite criterion to that for the trace offsets.

Some examples of the kind of diagnostics used to judge the desirability of a particular binning strategy are shown in Figs. 3.94 and 3.95. Fig. 3.94 shows the overall distribution of offsets and bin population for the whole line shown in Fig. 3.92. Some unevenness can be noted. Fig. 3.95 is a rather complex plot showing a detailed breakdown of various parameters for a particular bin, including the bin position on the processed line, the azimuthal distribution, the trace centroid and the offset distribution. Plots like this are normally only done at selected points on the line for diagnostic purposes.

Finally, Figure 3.96 shows the binning strategy chosen with the above points in mind. Note that a variable bin size was eventually used.

It was stated at the beginning of this section, that the non-trivial meaning of the phrase 'crooked line' referred to use of the surface spread of sources and receivers in order to derive information pertaining to the cross-line behaviour of particular horizons. In practice, this can be achieved by using several different, but parallel, processing lines, each with an appropriate

binning strategy. The result is several different output lines of lower fold (because the original field data is unchanged). Hence there is a trade-off between the signal-to-noise ratio of a given line and the amount of information relating to the cross-line direction. This can, however, yield vital information to the interpreter in the absence of a true 3-D survey.

3.6.3 3-D geometry and processing considerations

As was mentioned in the last section, the main difference between crooked line and 3-D geometry is one of

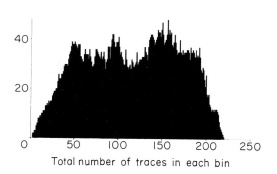

Total number of traces in each bin

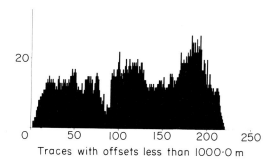

Traces with offsets less than 1000·0 m

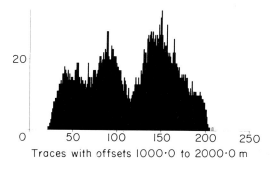

Traces with offsets 1000·0 to 2000·0 m

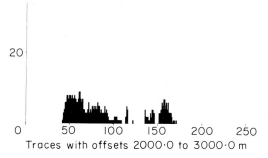

Traces with offsets 2000·0 to 3000·0 m

Fig. 3.94. Offset distribution and bin population diagnostics for the whole line.

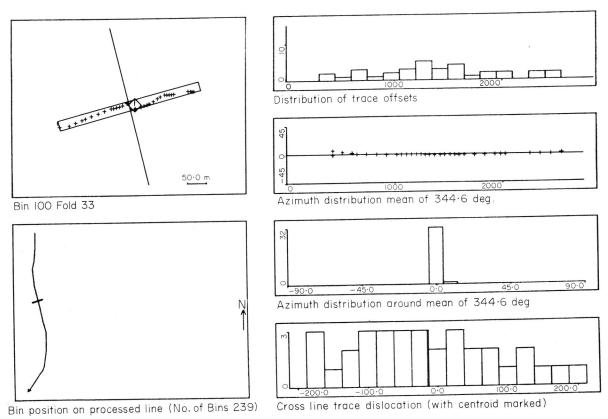

Distribution of trace offsets

Azimuth distribution mean of 344·6 deg

Azimuth distribution around mean of 344·6 deg

Bin 100 Fold 33

Bin position on processed line (No. of Bins 239)

Cross line trace dislocation (with centroid marked)

Fig. 3.95. Example diagnostics for a specific bin. Bin position on the line, azimuthal distribution, trace centroid and offset distribution are shown.

scale. By this it is meant that the size of the geometry database, which contains coordinate information for shot and receiver locations amongst other things, is much larger. The processing objectives in terms of bin population statistics are the same. Hence, it will suffice here simply to make the following points:

1   Regularity of acquisition is the key to good quality in data processing. Without regularity, 3-D processing inevitably involves a considerable amount of interpolation which generally damages the spatial resolution.
2   A parallel grid of straight acquisition lines is the optimal geometric configuration in general. Odd shapes such as squares and so on can be important for very specific objectives, but the inherent dramatic variation in bin population statistics such as total fold and azimuths, can severely hamper processing.

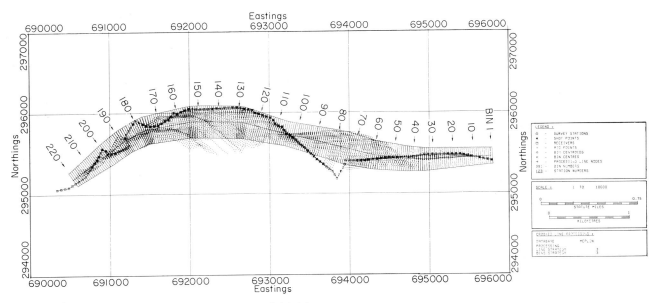

Fig. 3.96. Final 'binning strategy' for this line. A variable bin size has been used along the line in this case.

of equations such as (4.1.1) shows that knowledge of this alone is insufficient to provide a unique solution, and so for use in seismology (4.1.1) has to be approximated in some way.

4.1.2 The migrated time domain

These approximations will now be discussed in sufficient detail to bring out the existence of a third domain to compliment the two physical domains already defined above. This is known as the migrated time domain and is of paramount importance by virtue of the fact that almost all seismic interpretation is carried out in this domain. Regrettably, of the three domains, the migrated time domain is the only one which does not result from a seismic experiment and is therefore an odd choice in which to do interpretation, i.e. exercise geological and physical intuition. This choice is actually forced on the geologist and geophysicist because of a usually very inadequate knowledge of the velocity distribution within the earth.

It is reasonable to ask then, how the migrated time domain has come to occupy such a prominent position. The answer is that the practice of migration came first with the associated needs for economy, followed by the theory, rather later. The first implementations of migration involved the application of empiricism. As was described earlier, in order to undo the effects indicated on Fig. 4.2 it is necessary to sum amplitudes along hyperbolic trajectories whose curvature is governed by the medium velocity. This sum must then be placed at the apex or minimum time position of the hyperbola. Note that this correction naturally produces a time section as its output. This is, of course, highly desirable to the interpreting geophysicist, as any depth section produced by processing is anathema, due to the known inaccuracies involved in the estimation of velocities from surface seismic data and subsequent commitment to those velocities. When this was done in the early 1970s, people did not suspect the existence of the concept of depth migration and so the operation was known simply as migration. In today's parlance, it would of course be called a time migration.

Whilst this was being done, Jon Claerbout and his associates had set about the problem of migration by working directly with the differential wave equation (4.1.1). Using an ingenious coordinate transformation, the ill-posed nature was removed and the result cast into a form which could be used for migration. As this method had highly visible roots in the scalar wave equation, it came to be known as wave-equation migration and occurred in every other sentence uttered at the geophysical conventions of the mid 1970s. Claerbout's family of transformations were based on the following basic transformation.

First, without loss of generality, consider a two-dimensional earth for which

$$\frac{\partial P}{\partial y} \equiv 0 \qquad (4.1.2)$$

making $P = P(x, z, t)$ only.

Also assume that the medium velocity $V(x, z) = V_c$, a constant. Then consider the transformation

$$x' = x$$
$$z' = \frac{2z}{V_c} \qquad (4.1.3)$$
$$t' = t + \frac{2z}{V_c}$$

In this context, V_c is also known as the frame velocity.

Before looking at the effects on (4.1.1), it is worth making a few comments:
1 The new surface position coordinate x', is the same as the old.
2 The new vertical coordinate z' is now two-way travel time in common with previous seismic data processing stages.
3 The new time coordinate t' is weird. Trying to explain what this does used to be a popular way of whiling away the evenings at Stanford Exploration Project meetings. Its effect is to focus attention on the upcoming waves at the expense of the downgoing waves. If a negative sign had been used, the reverse effect of forward modelling or downward continuation is achieved. Migration is simply the backwards continuation of the upcoming waves until they are coincident with the reflectors from which they originated.

Using (4.1.2) and (4.1.3), (4.1.1) can be written

$$\frac{\partial^2 P'}{\partial z' \, \partial t'} = -\frac{V_c^2}{8}\frac{\partial^2 P'}{\partial x'^2} - \frac{1}{2}\frac{\partial^2 P'}{\partial z'^2} - \frac{3}{8}\frac{\partial^2 P'}{\partial t'^2} \qquad (4.1.4)$$

where $P'(x', z', t') = P(x, z, t)$.

Notice at this stage, no other approximations have been made. Equation (4.1.4) is exact for a two-dimensional earth of constant velocity.

In order to implement this equation, two controversial approximations were made.
1 The second term on the right of (4.1.4) was neglected. The argument for this was that the effects of the coordinate transformation guarantee that this term is small. This has the important effect of making the equation of first rather than second order in z'. This allows the equation to be solved using the surface recorded information $P'(x, 0, t)$ along with some appropriate boundary conditions in x'.

Note that simple neglect leads to the so-called 15 degree equation which only accurately propagates waves moving within 15 degrees of the vertical. The approximation can be iteratively refined in a way which still gives a first order equation in z', but with additional terms which allow accurate propagation at much wider angles. 45 degree algorithms are quite common.
2 The third term on the right of (4.1.4) was neglected for similar reasons to the second term. As it turns out, this has a crucial and unexpected effect.

The resulting equation was

$$\frac{\partial^2 P'}{\partial z' \, \partial t'} = -\frac{V_c^2}{8}\frac{\partial^2 P'}{\partial x'^2} \qquad (4.1.5)$$

So far, the development has taken account of constant velocities only. The equivalent development for the inhomogeneous case starts with the coordinate transformation.

$$x' = x$$

$$z' = \int_{b=0}^{z} \frac{2}{V(x, b)} \, db \tag{4.1.6}$$

$$t' = t + \int_{b=0}^{z} \frac{2}{V(x, b)} \, db$$

Note that z' is now a two-way travel time.

Mirroring the previous derivation and neglecting gradients of velocity, gives

$$\frac{\partial^2 P'}{\partial z' \, \partial t'} = -\frac{V'^2(x', z')}{8} \frac{\partial^2 P'}{\partial x'^2} \tag{4.1.7}$$

where $V'(x', z') = V(x, z)$.

This may be simply rewritten as

$$\frac{\partial}{\partial z'} \left[\frac{\partial P'}{\partial t'} \right] = -\frac{V'^2(x', z')}{8} \frac{\partial^2 P'}{\partial x'^2} \tag{4.1.8}$$

The migration at a particular z' is determined to be complete using the exploding reflector model due to Loewenthal *et al.* (1976). In this model, the earth is represented as a medium with half the velocity of the real earth. Hence a surface wave field recorded at time t, is coincident with the reflectors from whence it originated when the 'clock' has run back to zero, i.e. $t = 0$. Referring to equation (4.1.6), the migrated wave field then corresponds to $z' = t'$.

The complete boundary conditions for the resolution of equation (4.1.8) are then:

$P'(x', 0, t')$ is the surface-recorded wave field

$P'(x_0', z', t')$ is constrained to some value to form a side boundary condition. This value may be zero, an absorbing boundary condition, or some such

$P'(x_1', z', t')$ is similarly constrained to form the other side boundary condition

$P'(x', z', z')$ is the migrated time section.   (4.1.9)

Inspection of equation (4.1.8) shows that it can be obviously interpreted as a simple downward continuation in z', the two-way travel time, whereby the solution at $z' + dz'$ can be obtained from the solution at z', where dz' is the two-way travel time increment. As was mentioned earlier, the wave field is of course known at $z' = 0$. In practice, a value of between 5 and 15 times the sample interval is used and the method of finite differences employed.

In the meantime, following the elegant work of Newman (1975), Larner and Hatton (1976) had showed that the hyperbolic summation method with two additions was in fact a discrete approximation to the Kirchhoff integral solution to equation (4.1.1), and coined the name 'Kirchhoff summation' for this process. The two things which were necessary for consistency were:

1 A geometric spreading term corresponding in the two-dimensional case to

$$\frac{1}{(V_{RMS}^2 T)^{1/2}} \tag{4.1.10}$$

where V_{RMS} is the root mean square velocity. The reason for this is shown diagrammatically in Fig. 4.4. This result has considerable ramifications when the effects on noise are considered later.

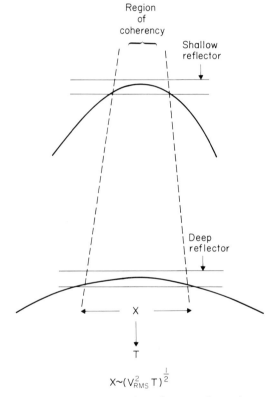

Fig. 4.4 The spread in the region of summation coherency or stationary phase with increasing time.

2 A spectral modification, corresponding to the differential term

$$\left[\frac{\partial}{\partial t'} \right]^{1/2} P(t) \tag{4.1.11}$$

which also appears in the 2-D Kirchhoff integral formulation. Analysing the effects of this differential operator on a simple harmonic function $\exp(2\pi i f t')$, shows that the spectrum must be modified by a filter which has the spectrum

$\pi/4$ phase shift

$(f)^{1/2}$ amplitude spectrum.   (4.1.12)

To see why this is the case, Newman's original work (1975) or Larner and Hatton (1976) should be consulted.

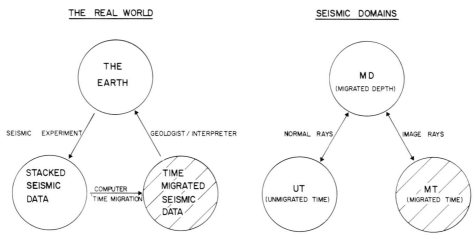

Fig. 4.5. A schematic illustrating the three domains of exploration seismology (from Parkes and Hatton, 1984).

To summarise the above work then, two migration methods based on the scalar wave equation were developed concurrently, both of which produced migrated time sections as a 'natural output'. Little did anybody suspect that Snell's law was absent from both.

4.1.3 Problems with velocity

The first glimmerings that something was seriously wrong came from the outstanding work of Hubral (1977), who showed that Kirchhoff summation did not, and could not, position images correctly if the velocity was allowed to vary laterally, even when the velocity was known. Hubral coined the term 'depth migration' to describe a secondary mapping to depth from the migrated time domain to correct for these deficiencies. This re-positioning amounts to depth conversion from migrated time by 'stretching' and 'bending' the migrated time traces along image rays, these being those rays which are perpendicular to the surface and which thereafter obey Snell's law. Larner *et al.* (1981) exploited this technique for weakly heterogeneous media, for which the two-fold action of migration, viz. the imaging or collapse of diffractions, and image-ray mapping, can be decomposed into two independent processes. In addition, Hatton *et al.* (1981) showed that finite-difference methods based on the differential form (4.1.6) had exactly the same deficiency and managed to trace the source of the problem to one of the approximations made in obtaining (4.1.5) from (4.1.4), viz. the neglect of the third term of (4.1.4) which is in effect where Snell's law and the image ray appear in this transformed wave equation. As will be seen in Section 4.2, the re-introduction of this term forces the natural output to be depth rather than two-way travel time.

Summarising, the three domains may therefore be re-defined as:

(a) *Unmigrated time.* The recorded echo-time domain in which no migration, effects are present.

(b) *Migrated time.* A time domain in which some aspects of migration, notably imaging (diffraction collapse), are present but in which Snell's law is absent. This domain is relatively insensitive to velocity error.

(c) *Migrated depth.* The geology domain. This domain is highly sensitive to velocity error.

These domains and their relationship are shown schematically in Fig. 4.5. Note that the unmigrated time and migrated depth domains are linked by normal incidence rays (rays which intersect the reflector of interest at right angles). Migrated depth and migrated time are similarly linked by image rays. Note also that the unmigrated time domain is here approximated by a stacked seismic section.

4.2 Natural migration coordinates

4.2.1 The meaning of natural coordinates

In this section, the idea of natural output coordinates for migration processes will be explored. Basically, natural coordinates allow simple physical interpretations to be made of mathematical equations. Unnatural coordinates preclude such interpretations.

As has already been seen, the natural interpretation of equation (4.1.8) is that it casts migration as a downward continuation in the variable z' defined by

$$z' = \int_{b=0}^{z} \frac{2}{V(x, b)} \, db \qquad (4.1.6)$$

which is the two-way travel time. This coordinate naturally emerges from the development, although a number of approximations are involved. Unfortunately, as was noted earlier, it turns out that Snell's law is absent from this formulation.

4.2.2 The migrated depth domain

The absence of Snell's law is of course a serious embarrassment to the average geophysicist working with wave equations, and what is life without a sense of Snell?

Consider therefore a more systematic treatment of equation (4.1.1) whereby unnecessary approximation is avoided. Using (4.1.2) and (4.1.3) in (4.1.1) directly, gives the following exact equation

$$\frac{\partial^2 P'}{\partial z' \, \partial t'} = -\frac{V_c^2}{8}\frac{\partial^2 P'}{\partial x'^2} - \frac{1}{2}\frac{\partial^2 P'}{\partial z'^2}$$
$$+ \frac{1}{2}\left(\frac{V_c^2}{4V^{2\prime}(x', z')} - 1\right)\frac{\partial^2 P'}{\partial z'^2} \qquad (4.2.1)$$

Again the approximation of directly or iteratively dropping the second term on the right-hand side of this equation is forced in order to obtain a soluble equation given the boundary conditions. This time however, the third term will be retained. Note that in this case, the downward continuation in z' is downward continuation in

$$z' = \frac{2z}{V_c} \qquad (4.2.2)$$

Although this has the dimension of time, it is, in fact, no more than a scaled depth coordinate, since V_c is constant. Hence this more systematic approach in which velocities are treated correctly, has led to a migration equation whose natural output coordinate is in depth. Solutions of this equation are therefore known as 'depth migrations'. It is important to note that for depth migrations, simple neglect of the second term is inadequate at all angles. See Hatton *et al.* (1981) for more details.

The next section will be devoted to velocity specification in migration and the sensitivity of various time and depth migrations to velocity uncertainty.

4.3 Migration velocities and velocity sensitivity

4.3.1 Introduction

In the past few years, understanding of the underlying algorithmic background of migration, thanks to the work of Berkhout (1980), Claerbout (1976), Gazdag (1978), Kjartansson (1979), Stolt (1978), and many others, has made enormous strides. Unfortunately, to a certain extent, this has dwarfed progress on what is the single most important issue facing the geophysicist desiring to use migration to simplify the interpretative task: that of the choice of velocity. Consequently, processing geophysicists seem to devote more energy to close scrutiny of the algorithm in use rather than the velocity actually being used.

Migration velocities tend to be treated in a cavalier fashion to say the least. For many years, the standard doctrine has been to use the stacking velocity, give or take an apparently random few per cent to account for the day of the week, the weather, or some other whim. The problem with stacking velocity quite apart from the fact that it refers to unmigrated time, is that it is so far divorced from reality as to have very little relationship with migration velocity in any realistic model.

4.3.2 Migration velocity

What then is migration velocity? To answer this question, recall the three domains as discussed in Section 4.1. The corresponding dependent variables in each domain are shown in Fig. 4.6. In an ideal world, the velocity would be known as a function of depth and an arbitrarily good depth migration algorithm used, before or after stack, 2-D or 3-D, depending on the degree of lateral heterogeneity. The result would be a near perfect depth or true migration. Hence, migration velocity for a depth migration is governed by the relevant set of dependent variables in the migrated depth domain:

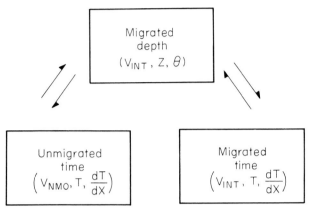

Fig. 4.6. The corresponding dependent variables appropriate to each domain.

interval velocities, depth and spatial dip of horizons. In the migrated time domain, a knowledge of the corresponding dependent variables, viz. internal velocities, time and time-dip, defines migration velocity for time migration algorithms. In practice of course, the following problems arise in migration velocity specification.

(a) The velocity model is known in unmigrated time rather than in migrated time. The key to progress here lies in the work of Hubral and Krey (1980) who describe relationships between the three domains. In particular, using the small offset approximation, they show how to map a velocity model from unmigrated time into migrated time. This is exemplified by Fig. 4.7a–b which shows the three domains for a given velocity model (actually specified in the migrated depth domain in this case). Here, a time migration based on a velocity model defined in unmigrated time, the solid horizons of Fig. 4.7b, will be significantly inaccurate in the very area which the interpreter would like to be clarified, that is around the unconformity. If it is assumed that velocity increases with increasing time in this model, the effect would be that the unconformity would be overmigrated.

(b) Time migration does not position events correctly in the presence of lateral velocity variations, even when the velocity field is known exactly. This has already been discussed in Section 4.1.3, leading to the concept of the image ray. Examples of image ray tracing for a number

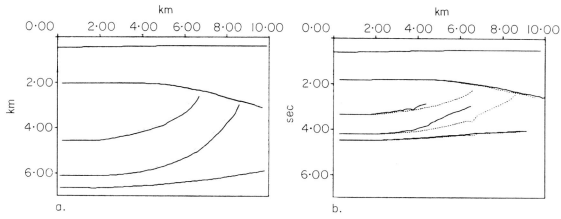

Fig. 4.7. The three domains for a particular unconformity model: (a) migrated depth, and (b) unmigrated time and migrated time (shown dotted).

of typical geological models are shown in Fig. 4.8a–e in comparison with a normal incidence ray model of selected layers in each model. Image rays have great diagnostic value in interpreting the effects of time migration as discussed by Hatton (1980). Two such effects are:

The diagnosis of positioning errors. Whenever an image ray departs from the vertical, the corresponding time migration has mispositioned events laterally. The image ray delineates the true lateral positions, presupposing the velocity model to be correct. On the models shown, this effect can be several hundred metres. In Fig. 4.7b, an examination of the image rays shows that the lateral variation is so severe that there is no map between migrated time and migrated depth for the deepest reflector. Comparison with the corresponding normal incidence ray model shows that ray paths are so confused for this reflector that a depth migration after stack and, more likely, before stack would be necessary to migrate the data from such a model.

The diagnosis of amplitude errors. Whenever image rays focus and defocus they are indicative of amplitude anomalies on the migrated time section caused solely by the time migration itself. When they focus or converge, they are trying to correct for amplitude deficiency on the migrated time section. When they defocus or diverge, the reverse is true. Hence, on Fig. 4.8a, the deepest reflector is likely to have a very uneven amplitude distribution on the migrated time section, even if it has a constant reflection coefficient. Incorrect interpretation of such amplitude anomalies can be very damaging, particularly in view of the faith geophysicists tend to have in the amplitude characteristics of the migrated time domain.

(c) 3-D effects are not properly handled by acquisition, let alone processing technique. This represents an area of intense current research. Some comments were made in Section 3.7 and all the ray-tracing algorithms used in this section are equally applicable to 2-D or 3-D, but the subject is generally beyond this text.

(d) Velocities derived from surface seismic measurements are statistically highly unreliable. In practice, the ray-tracing techniques discussed in (a) above are rather unsatisfactory as velocities must generally be specified in unmigrated time, as extracted from surface seismic data, and statistically these tend to be exceedingly erratic. Some kind of stabilisation is always necessary. Two possible approaches suggest themselves:

The use of well-log information. One of the unfortunate properties about unmigrated time is that, although the timing of reflectors is usually known extremely accurately, the velocity, as stated above, is usually hopeless. Much higher quality models can be obtained by using temporal information in the unmigrated time domain and velocity information from the migrated depth domain, i.e. a well.

The use of common sense. Although this property does not normally appear in theoretical textbooks, it can be of inestimable value. It is particularly necessary in the absence of well information of any kind, and generally involves model editing in the migrated depth domain following model computations using the unmigrated time domain parameters of NMO velocity, and the temporal shape of the model horizons.

Figs. 4.9a–d and 4.10a–d show just how complicated the velocity relationship between the three domains can be. Fig. 4.9 a shows a rather simple depth model consisting of conformable dipping layers. As shown in Fig. 4.9b, the interval velocities are constant in each layer and vary from 1500 m/s to 4500 m/s. The corresponding unmigrated and migrated time models are shown in Fig. 4.9c, and the NMO velocity variation in unmigrated time shown in Fig. 4.9d. In this case, comparing Fig. 4.9b and d, a rather uniform rise is observed in the NMO velocity as the horizons are traversed updip. As NMO velocity is normally identified with stacking velocity within reasonable bounds, this trend would also be observed in the stacking velocity. Given the smooth nature of the variation and reasonable S/N, it is likely that usage of the appropriate

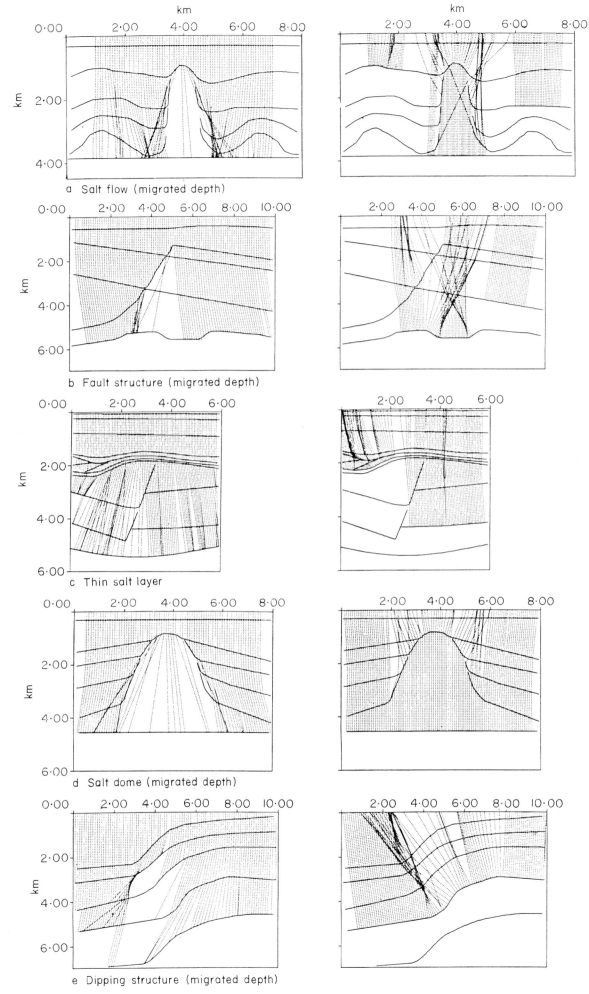

km

0·00   2·00   4·00   6·00   8·00

km

a Salt flow (migrated depth)

0·00   2·00   4·00   6·00   8·00   10·00

km

b Fault structure (migrated depth)

0·00   2·00   4·00   6·00

km

c Thin salt layer

0·00   2·00   4·00   6·00   8·00

km

d Salt dome (migrated depth)

0·00   2·00   4·00   6·00   8·00   10·00

km

e Dipping structure (migrated depth)

Fig. 4.8. Diagrams a–e show a variety of typical models comparing image ray trajectories to normal incidence ray trajectories from selected horizons.

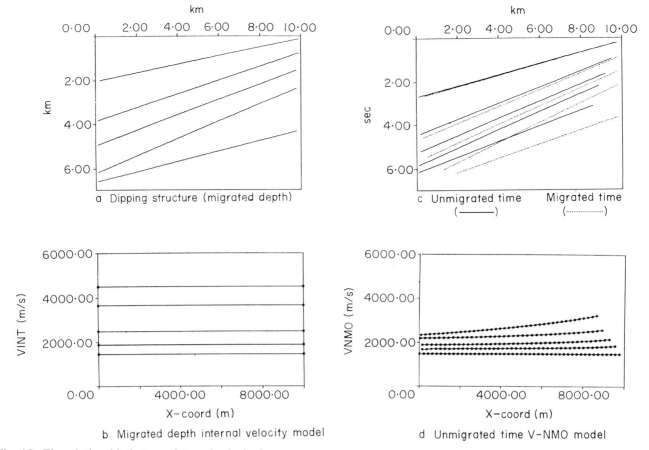

Fig. 4.9. The relationship between interval velocity in migrated depth and NMO velocity in unmigrated time for a simple model with conformable dipping layers. Dotted horizons are migrated time.

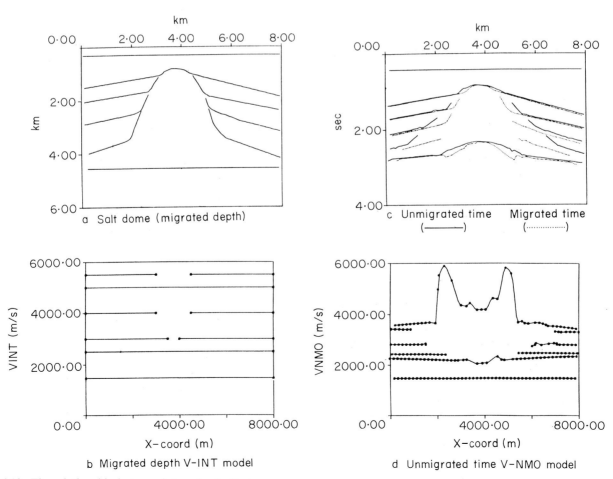

Fig. 4.10. The relationship between interval velocity in migrated depth and NMO velocity in unmigrated time for a more complex model consisting of a salt dome with onlapping sediments. Dotted horizons indicate migrated time.

inverse algorithm will produce a reasonable facsimile of the interval velocity distribution in the depth domain, Fig. 4.9b.

By contrast, Fig. 4.10a–d shows the velocity relationships for a slightly more complex, but very typical, depth model of a salt dome. A comparison of Fig. 4.10b and d shows a very much greater sensitivity to the effects of the model.

It is helpful to think of the depth model as the mapping function from one velocity domain to the other. In this case, starting with a knowledge of the velocity distribution shown in Fig. 4.10d in the unmigrated time domain, it is easy to imagine the difficulties arising in trying to reproduce the interval velocity model in migrated depth shown in Fig. 4.10b in the presence of noise. This inversion would certainly require considerable editing in migrated depth, the real domain, in order to produce a satisfactory velocity model for depth or time migration.

It should be noted that both of the above models take account of model curvature in the computation. In models like the latter, this is an important effect.

The issue of velocity uncertainty is of such importance in an industry which unfortunately lacks a systematic treatment of uncertainty in general, that some of its effects will now be explored in greater detail.

4.3.3 ... and sensitivity

To see the effects of uncertainty, consider Fig. 4.11 which is taken from Parkes and Hatton (1984). This shows the effects in the migrated time domain of perturbing the velocity in all layers of another salt dome model by plus or minus 20 per cent in the migrated depth domain, a not unreasonable figure in practice. Several points may be noted:
1 Even at this level of uncertainty, it is always better to use a model based on migrated time than on unmigrated time.
2 Errors are very small for small dip, as can be seen in the first layer.

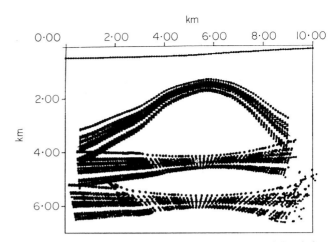

Fig. 4.12. Illustrating uncertainty zones in migrated depth for a perturbation in interval velocity of 20 per cent in the second layer only.

3 Cumulative errors are small for several overlying layers of small dip as can be seen by looking at the relatively flat layers under the anticline.

In comparison, Fig. 4.12 shows errors in the migrated depth domain when only one layer is subjected to a 20 per cent fluctuation. The effects are exceedingly disturbing and illustrate that migrated time is perhaps not such a bad domain to work in after all if uncertainty in velocity estimates is high. This will generally be the case in the absence of well information.

As a last example in this vein, Fig. 4.13 shows the fluctuations in image ray paths when the velocities in all layers are subjected to a 5 per cent perturbation. Such errors and their interpretation are also discussed by Larner et al. (1981).

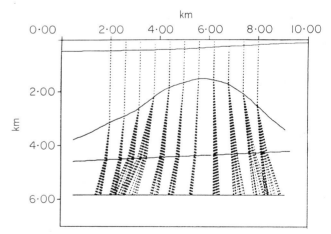

Fig. 4.13. Illustrating uncertainty in image ray positions for perturbations in interval velocity of 5 per cent in each layer.

4.4 Migration and noise

4.4.1 A Kirchhoff-summation formulation

The treatment of noise by migration algorithms has, unfortunately, often been neglected in the past at the expense of the treatment of signal, in spite of the fact that it can be of great importance. The effects will be discussed using the Kirchhoff-summation approach, as

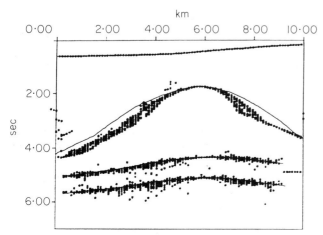

Fig. 4.11. Illustrating uncertainty zones in migrated time for perturbations in interval velocity of 20 per cent in each layer.

introduced in Section 4.1, in view of its simple visual-
isation. No reference to lateral heterogeneity will be
made in this section.

Recall from Section 4.1, that an intrinsic part of
migration according to the Kirchhoff-summation
method is to scale all data to compensate for geometric
spreading using equation (4.1.10). The reason for this is
that because the hyperbolic summation trajectory flattens with increasing time, the region of stationary phase
which governs the essential contribution to the summation, increases with time according to

$$(V_{RMS}^2 t)^{1/2}$$

Therefore later horizons of the same temporal duration
would contribute correspondingly more if the geometric
spreading term were not applied. However, considering
the summation of Gaussian noise, say, along the hyperbolic curve, the final summation is independent of the
curvature of the curve. The Kirchhoff integral knows
nothing of this of course, and the scaling is applied
anyway. Hence, after migration with any migration
method obeying the wave equation, an originally
uniform Gaussian noise background will have an amplitude variation given by

$$\frac{1}{(V_{RMS}^2 T)^{1/2}} \qquad (4.4.10)$$

So migration reduces the amplitude of background
noise without of course affecting the signal amplitudes.
This fact tends to take people by surprise although it is
not new and to the authors' knowledge was first
described by Gibson (1977).

Unfortunately, there is also a disadvantage. Consider
the summation trajectory in a high velocity medium.
The resulting trajectory is very nearly flat, especially at
later times, and so summation along it is very similar to
a lateral 'mix', which, as is well-known, emphasises the
lower wave numbers giving the data a smeared look
and greatly limits the lateral resolution. On very noisy
data of high velocity such as is acquired on land in
some parts of the world, this effect can be so severe as
to degrade the interpretability of migrated data compared with the unmigrated data. In a case like this, the
interpreter should interpret the unmigrated data and
ray-trace migrate the picked model as described in
Section 4.1. Wave theory does indeed have its limi-
tations in poor S/N.

The degree to which the amplitude effect indicated by
(4.4.10), and the resultant spectral smearing take place,
is governed by how accurately the algorithm in question
models the scalar wave equation. On the one hand,
Kirchhoff-summation and $f - k$ techniques numerically
simulate the scalar wave equation very accurately and
consequently produce the above effects faithfully,
tending to smear the data and change the spectral background. On the other hand, finite-difference techniques
do not generally model the wave equation well at
steeper dips and do not therefore change the spectral
background significantly. This probably explains their
great popularity in the industry as it is much easier for

the human brain to correlate two signals, before and
after migration, against a background noise field of the
same spectral qualities. If finite-difference techniques
capable of migrating steep dips accurately are used, they
promptly reproduce the noise smearing just as the
scalar wave equation says they must. Put simply, a
finite-difference time migration generally looks 'like'
the input stacked data to an interpreter.

Several predictions can be made from equation
(4.4.10).

1 The change in the amplitude of the noise background is not a function of velocity when the velocity is
constant. The amplitude simply varies as

$$\frac{1}{T^{1/2}}$$

which is, of course, less of a reduction than would occur
in a medium in which velocity increased with increasing
time. This disadvantage is offset by the fact that,
because the hyperbolae are less flat, there is less lateral
mixing. An important trade-off can be summarised as
follows: the more the amplitude of random noise is
reduced as a result of migration, the more the organisation of noise background is increased.

2 In a medium with a sudden large increase in velocity, a noticeable decrease in amplitude should occur at
the interface accompanied by a change in spectral character towards greater mixing. This can be seen in Figs.
4.14 and 4.15. Fig. 4.14 shows a simple zero-mean stationary Gaussian noise section, and Fig. 4.15 its migration using a finite-difference algorithm for a velocity
distribution of 2000 m/s in the first two seconds and
5000 m/s from two seconds to four seconds. Even for
the finite-difference algorithm which exhibits these
effects sub-optimally because of the steep-dip
restriction, the above characteristics can be clearly seen.

3 In three-dimensional migration, the equivalent
factor to (4.4.10) is

$$\frac{1}{(V_{RMS}^2 T)} \qquad (4.4.1)$$

This implies that Gaussian noise is further reduced in
three-dimensional migration than it would be in two-
dimensional migration. However, bearing in mind the
point made above, a correspondingly higher degree of
lateral mixing is to be expected. This factor may
account for apparently increased lateral smearing
observed in practice in three-dimensional migration.

4.4.2 2-D versus 3-D

It is worthwhile pursuing the relative reductions of
background noise achieved in practice in two- and
three-dimensional migration a little further.

Consider Fig. 4.16 which depicts a hyperbolic trajectory at some time in two-dimensions or a cross-section
of a hyperboloid in three-dimensions at the same time.
A dipping reflector is shown intersecting the trajectory.
The stationary phase region is assumed to contain M
traces and the migration aperture N traces. Consider a

Fig. 4.14. An unmigrated Gaussian noise section.

Fig. 4.15. The migrated equivalent of the noise section of Fig.
4.14 using a velocity of 2000 m/s above 2 seconds and
5000 m/s below 2 seconds. Note the reduction of amplitude
after 2 seconds.

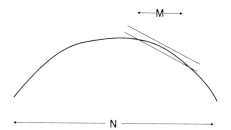

Fig. 4.16. Illustrating the size of the stationary phase region of M traces for a dipping reflector on a migration 'aperture' of N traces. Where M is proportional to the size of the stationary phase region, and N is proportional to the size of the effective algorithmic aperature.

simple summation along the trajectory. According to wave theory, signal is only present effectively in the region of stationary phase. Outside this region, only noise contributes. Misunderstanding this vital point is a common source of error in describing signal-to-noise ratios in migration. It has in the past led to extravagant statements to the effect that three-dimensional migration enjoys a much better signal-to-noise ratio than its two-dimensional equivalent because many thousands of input traces contribute to the output apicial trace. This is of course manifestly untrue.

Using Fig. 4.16, if without loss of generality the noise is assumed to be zero mean, unit variance Gaussian, and S the mean signal amplitude in the stationary phase region, the two-dimensional summation gives a signal-to-noise ratio of

$$S_{2D} \propto \frac{MS}{N^{1/2}} \qquad (4.4.2)$$

Note that the geometric spreading factor does not appear here as the signal-to-noise ratio is being considered.

The three-dimensional summation gives a signal-to-noise ratio of

$$S_{3D} \propto \left(\frac{\pi}{4}\right)^{1/2} \frac{M^2 S}{N} \qquad (4.4.3)$$

The ratio of the two is

$$\frac{S_{3D}}{S_{2D}} \propto \left(\frac{\pi}{4N}\right)^{1/2} M \qquad (4.4.4)$$

The respective sizes of M and N are a property of the particular algorithm used. The point is that (4.4.4) can be either side of one and is not an easy factor to calculate, hence, the issue of relative signal-to-noise ratios in two- and three-dimensional migration is not a simple one. Note once again, that the different geometric spreading factors do not intrude here as (4.4.4) is a ratio of ratios.

4.5 Migration and commutativity

4.5.1 The essence of commutativity

Two seismic data processes commute if the output is independent of the order in which they are applied.

Oddly enough, considering the large number of distinct processes to which most seismic data are subjected, there is very little in the literature which addresses this subject. In this section a few examples will be discussed, intended more to provoke thoughts than to present a formal theoretical basis of understanding. The essential point is that migration is not necessarily commutative. However, in practice, no serious problems appear to arise as a consequence.

4.5.2 Commutativity with filters

Consider first the simple example of convolution of two filters as described in Section 2.4. It is a well-known and easily provable fact that such filters do commute. Hence,

$$a * b = b * a \qquad (4.5.1)$$

It is probably this fact which causes many geophysicists to believe that filtering operations generally commute, including two-dimensional ones. In the case of migration, the fact that it is known to be an all-pass filter might add substance to this belief. In fact, by deriving the dispersion relation corresponding to equation (4.1.1) in a medium of constant velocity V_c, the following results:

$$\frac{k_x^2 + k_y^2 + k_z^2}{w^2} = \frac{1}{V_c^2} \qquad (4.5.2)$$

a constant, where k_x, k_y, k_z are the wave numbers in the various coordinate directions and w is the circular frequency. Hence, migration is all-pass only in the sense that it conserves the quantity on the left-hand side of equation (4.5.2). As will now be seen, this does not guarantee that migration commutes even with a single one-dimensional band-pass filter.

Consider the following two examples:

Single filter Fig. 4.17 shows what happens to two points on the same trace during migration with a constant velocity. As these points move up the summation hyperbolae, their temporal separation changes. Imagine then a filter which exactly removed the frequency corresponding to the separation before migration. If this filter were applied before migration, no information would contribute to the apex of the hyperbolae shown, as it would have been filtered out. If this filter were applied after the migration, it would have no effect on

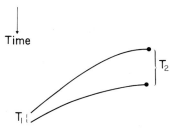

Fig. 4.17. Illustrating two half-diffraction curves for a medium of constant velocity V. Note the differential curvature.

the apical values as their new temporal separation is not affected by this filter. Hence migration does not commute with this single filter.

Multi-window filter Fig. 4.18 illustrates a dipping reflector in a section subject to a three zone filter, i.e. the filter properties are different in each zone. The zones do not vary with x, the profile direction. If, during migration, the dipping reflector moves from one zone to another, it is obvious that the filtering actually applied to the dipping reflector depends whether the filter is applied before or after migration. Hence, again migration does not commute, this time with a simple multiple zone filter.

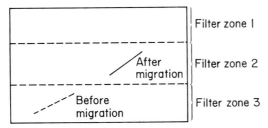

Fig. 4.18. A schematic indicating the relationship between a multi-zone filter and a migration. The dipping event lies in different zones before and after migration.

4.5.3 Commutativity with gain

The effects on commutativity of gain may also be anticipated. Quite simply, migration and gain do not commute. To see this, it is necessary to consider equation (4.1.11), corresponding to the term which is summed in the Kirchhoff integral.

$$\left[\frac{\partial}{\partial t'}\right]^{1/2} P(t) \qquad (4.1.11)$$

Suppose now that

$$P(t) = g(t) \cdot P'(t) \qquad (4.5.3)$$

i.e., P is a gained version of some original pressure function P', and g is the gaining function.

Substituting (4.5.3) into (4.1.11) and expanding, gives

$$\left[\frac{\partial}{\partial t'}\right]^{1/2} P(t) \qquad (4.5.4)$$

$$= \left\{ g(t) \cdot \left[\frac{\partial}{\partial t'}\right] P'(t) + P'(t) \cdot \left[\frac{\partial}{\partial t'}\right] g(t) \right\}^{1/2}$$

Inspection of equation (4.5.3) reveals that gain only commutes with migration if the second term on the right-hand side can be neglected with respect to first term. Hence commutativity holds if the gain is sufficiently slowly varying as a function of time. This is likely to be true in most cases, but not if a severe gain like an AGC is used.

Commutativity in the 3-D case can be analysed by removing the square root from the differential operator. Considering the effects on a simple harmonic function

as was done for equation (4.1.12), it can be seen that commutativity will be violated more at higher frequencies in the 3-D case than in the 2-D case.

Whilst these effects may seem slight, the topic of commutativity generally seems worthy of more thought particularly as modern trends in seismic data processing are seeking to extract more and more information from seismic data at higher and higher frequencies.

4.6 Migration and spatial aliassing

Spatial aliassing and its manifestation as dip reversal were discussed in Section 2.10. It is worthwhile returning to the subject briefly here, to describe the effects of migration on spatially aliassed data.

Approaching the subject from a slightly different viewpoint, Fig. 4.19 shows a single dipping reflector in a homogeneous medium of compressional velocity V. The reflector dips at angle α and the receivers are separated by a distance dx on the surface. According to the simple geometry of this diagram, all temporal frequencies in excess of

$$\frac{V}{2 \cdot dx \cdot \sin \alpha} \qquad (4.6.1)$$

are spatially aliassed. This corresponds exactly to the condition for dip-reversal discussed in Section 2.10.5.

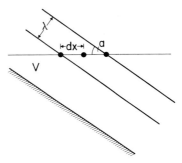

Fig. 4.19. Illustrating a wavefront from a dipping reflector which is just spatially aliassed at the surface receiver array.

There, dip-reversal was introduced using a two-dimensional time series with constant f, for which the dip (or k) was varied. In the context of migration, exactly the same thing occurs for constant k as f increases. Hence for all temporal frequencies greater than the limit given by equation (4.6.1), the dip is reversed and the migration consequently moves them in the wrong direction, leading to a virga of aliassed frequencies strewn somewhat indiscriminately around the locality of the dipping reflector.

From another point of view, migrating seismic data aliassed according to equation (4.6.1) band-limits reflectors, the aliassed frequencies simply appearing elsewhere. This effect can severely hamper resolution and is most prominant when low velocity, wide receiver separation and even modest spatial dips are combined. When these conditions occur, it may be better not to migrate, just as in the case mentioned earlier when discussing the influence of noise.

To summarise: for either high velocity, high noise; or low velocity, steep dips and wide receiver separation, it may be best to interpret the unmigrated data and migrate it ray-theoretically. Experience is the best judge.

The aliassing may not always be irremediable and two cases may be identified:

Benign aliassing For the situation described above, although the data are aliassed, the eye is quite able to interpret the dipping horizon. It is much more confused by the migration-induced virga. This seems to be a case where the eye is happy but the algorithm not. This is exactly the benign form of aliassing discussed by Larner *et al.* (1980), in their discussion of interpolation of data-sets aliassed according to equation (4.6.1). The aliassing is benign because it occurs in a multi-dimensional medium where there exists a rotation of this medium in which the data are not aliassed. In this case, the data can be interpolated in such a way that the above limit is raised by reducing dx, and correct migration is then possible.

Severe aliassing In this case, no such rotation exists as for example, where there are conflicting dips such as at an unconformity. The problem cannot be resolved algorithmically and it may be best not to migrate as described above.

Finally, note that the aliassing virga should not be confused with another, more regular kind of virga observed in shallow-dip finite-difference algorithms for unaliassed data and depicted in Fig. 4.20. This is simply due to positional errors within the algorithm itself, which increase with frequency for a given dip, and is unrelated to aliassing. The phenomena is also known as dispersion.

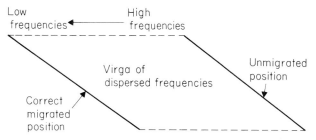

Fig. 4.20. This diagram shows the 'virga' of frequencies which occurs whenever a dipping reflector is migrated by a dispersive finite-difference algorithm.

4.7 Some useful migration formulae

It is common to find something like panic creeping over the average geophysicist when confronted by modern migration algorithmic development. Fortunately the basic rules haven't changed and the following simple formulae can be very useful in predicting some of the effects of time migration, whatever the algorithm used.

4.7.1 Approximate position after migration

This can be calculated by using the equation for the summation trajectory in the Kirchhoff summation. This is

$$T^2 = T_0^2 + \frac{X^2}{V_{RMS}^2} \qquad (4.7.1)$$

where T_0 represents the time at the apex of the hyperbola, X the horizontal offset from the apex corresponding to time T, and V_{RMS} is the RMS velocity at the apex position. In general, the latter will not be known and should be taken as the value at the unmigrated position.

Taking the derivative of this with respect to X and rewriting, gives

$$X = V_{RMS}^2 \, T \, \frac{dT}{dX} \qquad (4.7.2)$$

Equations (4.7.1) and (4.7.2) between them are enough to determine an approximate migrated position. For example: suppose $T = 2.0$ seconds, $V = 2000$ m/s, and $dT/dX = 5$ msec per trace with a trace separation of 25 metres. Then equation (4.7.2) gives

$$X = (2000)^2 2 \, \frac{.005}{25}$$

$$= 1600 \text{ metres or 64 traces.}$$

Using this in equation (4.7.1), gives

$$T_0 = \left(4 - \frac{1600^2}{2000^2} \right)^{1/2}$$

$$= 1.833 \text{ seconds.}$$

Hence the point at 2 seconds will move approximately 64 traces in the direction of dip, to two-way time 1.833 seconds.

4.7.2 Dip before and after migration

Simple geometry as indicated in Fig. 4.21 shows that the dip of a single layer on an unmigrated section and a migrated section is given by

Unmigrated dip $\quad \left(\dfrac{dT}{dX} \right)_u = \dfrac{2 \sin (a)}{V}$

Migrated dip $\quad \left(\dfrac{dT}{dX} \right)_m = \dfrac{2 \tan (a)}{V}$

where a is the spatial dip of the layer and the medium is homogeneous with velocity V.

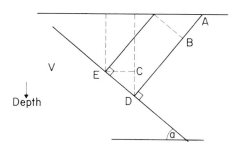

Fig. 4.21. The nomenclature used to develop the simple migration formulae of Section 4.7.

Note that these expressions can also be used inversely to calculate approximately the real spatial dip from the unmigrated or migrated time dips.

For example, a real spatial dip of 10 degrees in a medium of velocity 2000 m/s and trace separation of 25 metres, will have unmigrated and migrated time dips of 4.3 and 4.4 msec/trace respectively. This closeness is of course a result of the similarity of sin and tan functions for small 'a'. For 30 degrees, however, the corresponding values are 12.5 and 14.4 msec/trace.

4.8 Migration before stack

Nothing has so far been said on the subject of migration before stack. In essence, the technique can be viewed as migration of each contributing constant offset section using the appropriate medium velocity. This effectively removes the cosine dependence of the stacking velocity on dip as exemplified by equation (3.3.10), forces spatial alignment and allows events of different dip at the same unmigrated time to stack correctly with the same velocity. This technique has been around for as long as migration itself, but is rarely used in practice due to its much higher cost (migration after stack cost times the fold of stack in its fullest form). As computers become more powerful, the extra cost becomes less of a disadvantage.

The important question however is, 'Is it worth it?' On the face of it, the technique would seem to be of considerable importance, especially as the industry is forever searching for higher frequency data and therefore better resolution. In practice, however, the answer seems to be generally no. It has certainly been tried enough in the past, but almost always with disappointing results vis-à-vis the cost. Why should this be so? The answer, as Hatton et al. (1981) discuss, may well be that any significant lateral velocity variations will render time migration before stack a pointless exercise, necessitating depth migration before stack as described by Schultz and Sherwood (1980). If the medium velocity is highly homogeneous, there exist much cheaper alternatives, such as discussed in the excellent review by Hood (1981). Nevertheless, the technique retains the faith of many processing geophysicists for reasons which are difficult to understand and its usage will no doubt continue. In addition to Hood (1981) another general review article which is strongly recommended is Hosken (1981), and also the texts by Robinson (1983), Berkhout (1984b) and Claerbout (1985).

Chapter 5

Inverse Theory and Applications

5.1 Introduction

Many inverse solutions exist for particular geophysical problems or types of problems. Two familiar examples are the Herglotz–Weichart equation, which enables seismic velocities within a laterally homogeneous earth to be determined from global travel time data, and Dix's equation which relates the velocities within a stack of horizontal layers to the travel times of reflected waves from the interfaces between the layers. 'Inversion' simply means that the velocities can be determined directly from the observed travel times. The corresponding forward solution would permit theoretical travel times to be calculated for any specified velocity model.

In all the discussion of differing methods and applications that follows, it will be assumed that a theoretical relationship has been developed that adequately predicts the values of the experimental data. In other words, the 'forward problem' is assumed to be solved. A number of inverse methods are described that are generally applicable to a wide variety of geophysical problems. All the examples of the application of these methods have been taken from the field of seismology. This is simply because the book is fundamentally concerned with seismic data processing. It would have been just as easy to include examples from any other field of geophysical exploration.

5.2 Non-uniqueness

The question of uncertainty in parameter estimates will be a dominant theme in this chapter and a few introductory comments on the subject are necessary.

Whenever an earth model is determined, by whatever means, it is crucial to know whether it is unique. If it is not, then it is possible that it only represents one of many (possibly infinitely many) very different models that satisfy the data. In other words, the data do not apply any useful constraints on the particular solution being sought. However, if the acceptable models all fall within some narrow bounds, then the determination of these bounds on the physical properties may provide some useful geological information. Note that the statement that a certain set of data can be satisfied by an infinite number of models does not imply that the models are infinitely variable over all possible model parameter values. An infinite number of models can exist within narrow parameter bounds.

Ideally, every geophysical model should include some estimate of the non-uniqueness bounds on the model parameters. Whenever a model is published without such estimates, the author is effectively saying that the non-uniqueness of the model is trivially small, at least for the purposes of the argument being presented. This may or may not be the case. However, it is certainly true that the non-uniqueness of a particular inverse problem can be effectively removed by careful construction of the type of model sought.

For example, gravity data are inherently non-unique. The equivalent layer theorem of potential theory states that the gravity potential at any point outside a surface which is due to matter inside the surface is the same as would be produced by a layer of matter of surface density:

$$\delta = \frac{1}{4\pi} \frac{\partial v}{\partial n}$$

spread over the surface, where v is the potential of the whole of the attracting matter and n represents the outward normal direction from the surface. Since the shape of the surface enclosing the matter is arbitrary, this theorem gives a clear indication of the range of distributions of matter that could satisfy any one gravity potential function. This does not mean that gravity data is useless. But it does mean that initial constraints have to be imposed upon the distribution of the matter in order that potentially unique models can be obtained. If one is confident enough to specify that the body must be a sphere, or a cylinder, or rectangular shaped, then very simple inverse expressions can be deduced that lead to direct, unique values of the dimensions of the simple bodies. But these values are believable only to the extent that the original imposed assumption about the general shape can be believed.

The phrase 'potentially unique' has been used above since no consideration has yet been given to noise in the data. A certain degree of non-uniqueness in solutions will always result from uncertainties in the observed data. But for the time being, only non-uniqueness resulting from perfect, noise-free data is being discussed.

It is important to appreciate that non-uniqueness bounds quoted for a particular model are critically dependent upon any assumptions incorporated in the inverse procedure. As an obvious example, any estimate of error in the parameters of a plane-layered earth model deduced from seismic refraction data would only be meaningful if it were clearly stated whether or not the inversion scheme involved the assumption of horizontal layering. However, sometimes assumptions incorporated into an inverse procedure are very easy to overlook, as will be illustrated in the next section.

5.3 Non-linear methods

All the inversion methods described in Sections 5.4 to 5.12 are based on the assumption that the equation relating observational data values to model parameters is linear. This is not nearly as restrictive as it might first appear since solutions to non-linear problems can be achieved via a step-wise sequence of linear approximations (see Section 5.5). However, there are a family of inversion methods which do not require any assumptions concerning the linearity of the governing equations. These will now be briefly considered.

The simplest method of deducing model parameters that satisfy some observed dataset is to resort to trial and error. While this is a perfectly respectable procedure, it is usually extremely laborious, and the labour is increased to unmanageable proportions if non-uniqueness bounds on the parameters are also sought. 'Monte Carlo inversion' is simply trial and error carried out by a computer which has a great deal more patience than the average protein-driven experimentalist. The computer programme consists of a routine for calculating theoretical data values for any model of earth physical parameters. These theoretical data are then compared with the observed data. If the fit passes some acceptance criteria, the model is accepted. If not it is rejected and forgotten. The computer may carry out this procedure literally millions of times and great care is exercised in making the automatic generation of models to be tested a perfectly random operation. It is hoped that this removes all human bias from the solutions. The final solution consists of non-uniqueness bounds which are defined by the superposition of all acceptable models.

In a variant of this method, known as 'Hedgehog inversion', parameter space is searched in a more systematic manner. Random models are generated and tested as in the Monte Carlo procedure. However, once an acceptable model has been discovered, subsequent models are not randomly generated. The method tests the nearest neighbours to this successful point in parameter space until a successful region within the vicinity of the original successful point has been defined. Normally, the procedure is then to revert to a random search and repeat the process if another successful model is discovered within some other region of parameter space.

A problem with random search procedures is that one is never completely certain that a sufficient number of trials have been attempted. It must be appreciated that for a model with only ten parameters where each parameter is permitted ten possible values, there exist 10^{10} potentially acceptable models. In practice, this figure would probably be greatly reduced by constraints imposed upon the form of the solutions; for example, the smoothness of the model or whether low velocity

layers were permitted in a velocity-depth model and, if so, how many.

If a single model is desired which minimises some acceptance criterion, normally referred to as the 'objective function', then a whole range of methods are available. A typical objective function might be the sum of the squares of the differences between observed and calculated data values. Imagine, if you can, a surface in n dimensional space defined by the value of the objective function as a function of the $(n-1)$ parameters. If the mind boggles, think of a surface in three-dimensional space where the objective function is dependent on only two model parameters. The aim is to find one's way to the middle of the deepest valley in the objective function surface as efficiently as possible. The minimum is approached via a series of increments in the parameter values. The various methods differ in the manner in which the size and direction of each parameter increment is determined. For example, 'gradient methods' find the most direct way down hill by calculating partial derivatives of the objective function with respect to the parameters. For further explanation and discussion, the reader is referred to Adby and Dempster (1974) and Greig (1980).

Regardless of the algorithm chosen, one must always be mindful of the danger of ending up in a local, rather than the global minimum (Fig. 5.1). It is generally good practice to attempt the minimisation two or three times starting on each occasion from a different point in parameter space. Alternatively, the region within the vicinity of a minimum can be explored, as in Hedgehog inversion, to discover whether a deeper valley is not lurking beyond a nearby ridge.

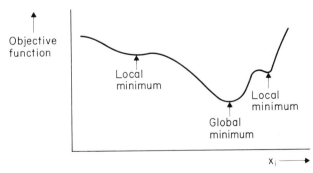

Fig. 5.1. The objective function in n dimensional parameter space may have any number of local minima into which an iterative algorithm might fall whilst searching for the global minimum.

One significant advantage of these procedures compared to other inversion schemes described below, is that there is effectively no limit to the complexity of the model constraints that can be imposed. One simply constructs a more elaborate model generating subroutine and pays the price in computer run time. The main disadvantage of all non-linear inversion methods is that with increasing numbers of parameters they rapidly become quite unacceptably time-consuming. It is for this reason that the methods are not currently widely used in seismic applications.

Non-uniqueness bounds defined by the superposition of all the acceptable models obtained from a Monte Carlo or Hedgehog inversion will be very dependent upon any *a priori* constraints imposed on the models. Fig. 5.2a shows a suite of upper mantle shear velocity models derived from a Monte Carlo inversion of global seismic travel time data (Worthington *et al.*, 1974). It would appear that the suite of models provide confidence limits on, say, the shear velocity at 400 km depth in the mantle. It should be noted that one key assumption in the production of Fig. 5.2a was that only two low-velocity channels were permitted. However, there is another hidden smoothing assumption. The bounds are also critically dependent upon the spacing of the grid upon which the random earth models are constructed.

Consider two extreme cases. If the grid spacing was extremely fine, it would be possible to generate models like the one shown in Fig. 5.2b. These models may satisfy the data since the anomalous thin layer would be too fine to be detectable, given only a finite set of data. Non-uniqueness bounds based on this grid spacing would be so broad as to be worthless. At the other extreme, a grid spacing could be arbitrarily chosen that was too coarse, so that the computer could not produce any models with sufficient structure in them to satisfy the data.

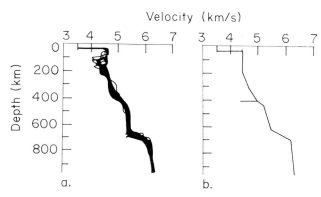

Fig. 5.2. (a) A set of upper mantle shear velocity models generated with a Monte Carlo inversion procedure; (b) one upper mantle model that would satisfy the acceptance criteria described in the text.

Somewhere between these two extremes there exists a model that is just complex enough to satisfy the data but simple enough (i.e., built on a sufficiently coarse grid) so that its non-uniqueness bounds are essentially zero. One could call the search for such a model the 'method of maximum simplicity'.

However, even if a solution is expressed as a model representing maximum simplicity, it is still necessary to specify what is meant by 'simple'. Generally, some definition of smoothness, s, is involved. For example, minimise

$$s = \sum_{i=1}^{n} (x_i - x_{i-1})^2$$

Alternatively, the model might consist of constant values of the parameters over specified depth intervals (a step model).

From the contents of Sections 5.2 and 5.3, the reader will appreciate that the concept of non-uniqueness bounds on the results of inversion require careful definition. This problem will be taken a stage further when, in a later section, it will be admitted that experimental data are subject to error.

5.4 Matrix formulation of the inverse problem

Certain geophysical problems are readily formulated in terms of a set of linear equations:

$$y = Ax \qquad (5.4.1)$$

where y represents a column vector of observed data values, x represents a column vector of model parameter values and A is a $(m \times n)$ matrix of co-efficients that relates the m observed data to the n model parameters. The inverse problem is the solution for x of equation (5.4.1). Essentially this is achieved by pre-multiplying equation (5.4.1) by a matrix operator H, such that:

$$HA \approx I$$

Then

$$Hy = HAx \approx \hat{x} \qquad (5.4.1a)$$

If HA is not an identity matrix, then the parameters of the model $\hat{x}$ are some weighted average of the values of the true solution x. For example,

$$\hat{x}_6 = \sum_{i=1}^{n} r_{6i} x_i \qquad (5.4.2)$$

where r_{6i} are the elements of the sixth row of matrix HA. Once some optimal value of H has been obtained, then the ith row of HA represents the 'resolution function' associated with parameter i. In general, these are bell-shaped functions that peak about the diagonal element (Fig. 5.3).

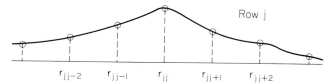

Fig. 5.3. A pictorial representation of one row of a resolution matrix.

If x represents some earth parameter that changes with depth, then the breadth of this function represents the resolving length of the data over some depth interval in the earth. It can be a most useful measure of the non-uniqueness of a model.

The problem, of course, is how to determine H; a question that will be considered in some detail in later sections. For the time being, consider only how various seismic problems can be formulated in terms of equation (5.4.1).

One example has already been described. In Section 2.8.1 it was shown how the discrete linear convolution

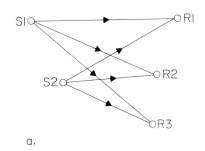

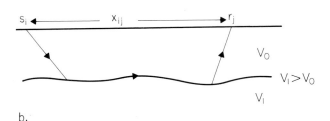

b.

Fig. 5.4. A simple delay-time experiment. (a) Two shots $S1$ and $S2$ are recorded at positions $R1$, $R2$ and $R3$. (b) The geological structure consists of a layer, velocity V_0 over a medium with velocity V_1, $V_0 < V_1$. The undulations of the interface are small compared to the thickness of the layer.

of two time series can be expressed as a matrix equation. The corresponding inverse problem is the process of deconvolution. (See Section 5.8 for details.)

Consider now the refraction seismology delay-time experiment (Fig. 5.4). If the earth consists of a layer of velocity V_0 over a sub-stratum of velocity V_1 (where $V_0 < V_1$), then the travel time between a shot and a receiver on the surface t_{ij} is

$$t_{ij} = \frac{x_{ij}}{V_1} + \delta_i + \gamma_j \tag{5.4.3}$$

where x_{ij} is the horizontal distance between shot and receiver and δ_i and γ_j are the delay times associated with the ith shot and the jth receiver positions. Consequently, if there were two shots each recorded at three receiver positions, the matrix formulation of the problem could be as follows:

$$\begin{pmatrix} t_{11} \\ t_{12} \\ t_{13} \\ t_{21} \\ t_{22} \\ t_{23} \end{pmatrix} = \begin{bmatrix} 1 & 0 & 1 & 0 & 0 & x_{11} \\ 1 & 0 & 0 & 1 & 0 & x_{12} \\ 1 & 0 & 0 & 0 & 1 & x_{13} \\ 0 & 1 & 1 & 0 & 0 & x_{21} \\ 0 & 1 & 0 & 1 & 0 & x_{22} \\ 0 & 1 & 0 & 0 & 1 & x_{23} \end{bmatrix} \begin{pmatrix} \delta_1 \\ \delta_2 \\ \gamma_1 \\ \gamma_2 \\ \gamma_3 \\ 1/V_1 \end{pmatrix} \tag{5.4.4}$$

The inverse of this particular matrix equation will be described in Section 5.6.

A number of problems can be formulated as an integral equation of the form:

$$d_i = \int_0^z K_i(z)p(z)\,dz \tag{5.4.5}$$

where d_i are a finite set of data, $p(z)$ represents some continuous function to be determined, for example the velocity or density as a function of depth in a laterally homogeneous earth, and d_i and $p(z)$ are related by the functions $K_i(z)$. If the integral is approximated in a com-

puter by a summation, with Δz set at some suitably small value, then equation (5.4.5) has the form $y = Ax$, where $y = (d_i)$ and $x = p(m\Delta z)$. However, in general there will be vastly more unknowns than equations. In the limit as $\Delta z \to 0$ and $p(m\Delta z) \to$ [the continuous function $p(z)$], the number of unknowns becomes infinite. It is therefore necessary to simplify the earth model $p(z)$ in some way. In matrix terms

$$x = Wx' \tag{5.4.6}$$

where W is $m \times n$ matrix which transforms the vector x of length m (where m is normally very large) into a vector x' of length $n(n \ll m)$. Consider two examples of W:

1   The earth is assumed to consist of only five layers within which the earth parameter is constant.

$$\begin{pmatrix} x_1 \\ x_2 \\ \vdots \\ \vdots \\ x_j \\ x_{j+1} \\ x_{j+2} \\ \vdots \\ \vdots \end{pmatrix} = \begin{bmatrix} 1 & 0 & 0 & 0 & 0 \\ 1 & 0 & 0 & 0 & 0 \\ & & \vdots & & \\ 1 & 0 & 0 & 0 & 0 \\ 0 & 1 & 0 & 0 & 0 \\ 0 & 1 & 0 & 0 & 0 \\ & & \vdots & & \\ & & \text{etc.} & & \end{bmatrix} \begin{pmatrix} x_1' \\ x_2' \\ x_3' \\ x_4' \\ x_5' \end{pmatrix} \tag{5.4.7}$$

2   The earth parameters are assumed to vary linearly between a finite number of depth points at which the earth parameter is to be determined.

$$\begin{pmatrix} \vdots \\ x_{j-2} \\ x_{j-1} \\ x_j \\ x_{j+1} \\ x_{j+2} \\ \vdots \\ \vdots \end{pmatrix} = \begin{bmatrix} & & \vdots & & \\ 0 & 4 & 6 & 0 & 0 \\ 0 & 2 & 8 & 0 & 0 \\ 0 & 0 & 1 & 0 & 0 \\ 0 & 0 & 8 & 2 & 0 \\ 0 & 0 & 6 & 4 & 0 \\ 0 & 0 & 4 & 6 & 0 \\ 0 & 0 & 2 & 8 & 0 \\ 0 & 0 & 0 & 1 & 0 \\ & & \vdots & & \end{bmatrix} \begin{pmatrix} x_1' \\ x_2' \\ x_3' \\ x_4' \\ x_5' \end{pmatrix} \tag{5.4.8}$$

Substituting equation (5.4.6) into equation (5.4.1)

$$y = AWx' = A'x' \tag{5.4.9}$$

where the ith row of the matrix A' is the numerical solution of the equation

$$A'_{ij} = \int_0^z K_i(z)W_j(z)\,dz \approx \sum_k K_{ik}W_{kj} \tag{5.4.10}$$

where $W_j(z)$ is represented by the jth column of the matrix W.

5.5   Linearisation

Unfortunately, it has to be admitted that most equations relating geophysical observations to earth parameters are not linear. So before the impressive battalions of linear algebra can be of general use, it is necessary to apply some procedure that reduces any problem to a linear form.

Assume that it is possible to make some intelligent initial guess as to the solution $(\hat{x}_1, \hat{x}_2, \ldots, \hat{x}_n)$. From the solution of the forward problem, theoretical data values are obtained:

$$\hat{y}_i = f(\hat{x}_1, \hat{x}_2, \ldots, \hat{x}_n) \qquad i = 1, m \qquad (5.5.1)$$

The function y is expanded about the point $(\hat{x}_1, \hat{x}_2, \ldots, \hat{x}_n)$ in parameter space

$$y_i(\hat{x}_1 + \delta \hat{x}_1, \hat{x}_2 + \delta \hat{x}_2, \ldots, \hat{x}_n + \delta x_n)$$

$$= \hat{y}_i + \frac{\partial f_i}{\partial x_1} \delta x_1$$

$$+ \frac{\partial f_i}{\partial x_2} \delta x_2 + \cdots + \frac{\partial f_i}{\partial x_n} \delta x_n$$

$$+ \text{ higher order terms} \qquad i = 1, m \qquad (5.5.2)$$

Provided δx_i, for all i, are small, the higher order terms can be ignored. In which case,

$$y_i - \hat{y}_i = \Delta y_i$$

$$= \frac{\partial f_i}{\partial x_1} \delta x_1 + \frac{\partial f_i}{\partial x_2} \delta x_2 + \cdots + \frac{\partial f_i}{\partial x_n} \delta x_n$$

$$i = 1, m \qquad (5.5.3)$$

$$Y = AX$$

where Y is the vector of differences between the observed data and the calculated data for the starting model; X is the vector of changes in the earth parameters which result in this change in the data; and A is the matrix of partial derivatives which represent the amount that the data is affected by a small change in any one of the earth parameters. Thus a set of linear equations is established. The final model is

$$(\hat{x} + X) \qquad (5.5.4)$$

This works well provided that the higher order terms in the expansion really are negligibly small. The larger the value of X, the more likely it is that the linear assumption will lead to error. In practice, it is usually necessary to iterate. Hence the model defined in equation (5.5.4) becomes the new x. A new set of partial derivatives are determined for the elements of the new A matrix, and so a new X is determined. This process can be repeated as many times as is required for the solution to converge to some stable value.

Two problems that frequently arise concern convergence and computing time. The question of convergence will be considered in more detail in Section 5.8. The calculation of the A matrix, which must be accomplished at each iteration, can be very computationally expensive. For each element

$$\frac{\partial f_i}{\partial x_j} = \frac{\begin{array}{c} f_i(\hat{x}_1, \hat{x}_2, \ldots, \hat{x}_j + \delta x_j/2, \hat{x}_{j+1}, \ldots) \\ - f_i(\hat{x}_1, \ldots, \hat{x}_j - \delta x_j/2, \hat{x}_{j+1}, \ldots) \end{array}}{\delta x_j}$$

$$(5.5.5)$$

which implies the running of the forward problem algorithm twice for each column of the A matrix.

5.6 Generalised matrix inversion

In this section, the mathematical content has been kept to an absolute minimum. A number of excellent references exist which describe the theoretical basis of the procedure known as generalised matrix inversion (GMI). The following are particularly recommended: Aki and Richards (1980), Lanczos (1961) and Jackson (1972). The emphasis of this section will be on the practical application of GMI software to geophysical problems. It is worth pointing out that virtually any mainframe computer will have, within its subroutine library, the necessary software to carry out any operation described in this chapter.

Consider the equation

$$y = Ax \qquad (5.6.1)$$

in which A is a square, symmetric matrix. The vectors y and x can be transformed into new vectors Y and X by the matrix operator V. Thus

$$y = VY, \, x = VX$$

and so

$$VY = AVX$$

If V is orthogonal,

$$Y = V^T AVX = \Lambda X \qquad (5.6.2)$$

It will be assumed that the reader appreciates that Λ is a diagonal matrix with the eigenvalues of A as the elements of the main diagonal, and V is a matrix whose columns are the normalised eigenvectors of A. From an examination of the eigenvalues of A, one can tell at a glance whether a reliable estimate of x can be obtained from y and A^{-1}.

Since

$$x = V \begin{bmatrix} \frac{1}{\lambda_1} & & & & \\ & \frac{1}{\lambda_2} & & 0 & \\ & & \ddots & & \\ & 0 & & \ddots & \\ & & & & \frac{1}{\lambda_n} \end{bmatrix} Y \qquad (5.6.3)$$

any very small values of λ (very large values of $1/\lambda$) will have a dominant effect on the estimate of x. This is of no consequence unless it is admitted that a small number is likely to be predominantly composed of noise and/or numerical error. In which case the estimate of x will be highly unreliable. The question of how small is unacceptably small will be considered in due course.

The above procedure can be extended to the more general case of a non-symmetric, non-square ($n \times m$) matrix by transforming x and y with different orthogonal matrices $V(m \times m)$ and $U(n \times n)$. Thus $x = V\alpha$, $y = U\beta$.

$$\beta = U^T AV\alpha = \Lambda\alpha \qquad (5.6.4)$$

where Λ is a diagonal matrix, with at most $q = \min(n, m)$ non-zero diagonal elements (see Appendix A for details). Thus $A = U\Lambda V^T$.

For an initial understanding of the method, the reader need not be aware of precisely how U, V and Λ are determined. (Discussion of this point is covered in Appendix A.) The factoring of A into U, V and Λ is referred to as singular value decomposition.

The main objective of the present discussion is the definition of an acceptable inverse operator H, as introduced in Section 5.4. A promising candidate would seem to be:

$$H = V\Lambda^{-1}U^T \qquad (5.6.5)$$

since V and U are orthogonal square matrices

$$HA = V\Lambda^{-1}U^T U\Lambda V^T = I$$

However, as in the simple square, symmetric case, problems arise when the main diagonal elements of Λ are zero or very small.

Suppose that a geophysical problem has been set up in a manner described in Section 5.4 and the number of equations (data values) equals or exceeds the number of unknowns (parameter values). The matrix A is decomposed into its U, V, Λ components, shown schematically below. Assume that there are one or more zero or effectively zero main diagonal elements (singular values) in Λ so Λ^{-1} cannot be formed. Columns in U and V labelled U_p and V_p can be associated with the non-zero singular values.

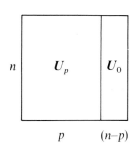

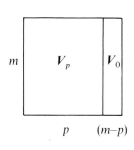

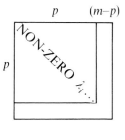

Since U and V are orthogonal, it follows that:

$$U_p^T U_p = I \text{ and } U_0^T U_0 = I \qquad (5.6.6)$$

but $U_p^T U_0 = U_0^T U_p = 0$ since different rows are being multiplied together. The same is true for matrix V. The transformations from (x, y) to (α, β) now become:

$$y = U_p\beta_p + U_0\beta_0 \qquad (5.6.7)$$

$$x = V_p\alpha_p + V_0\alpha_0 \qquad (5.6.8)$$

The generalised inverse operator H is defined as:

$$V_p\Lambda_p^{-1}U_p^T \qquad (5.6.9)$$

The first obvious fact to appreciate about this operator is that it always exists since the elements in Λ^{-1} which would cause trouble have been eliminated. The estimate of x achieved with this inverse is $\hat{x}$,

$$\hat{x} = Hy \qquad (5.6.9a)$$

Substituting (5.6.7), (5.6.8) and (5.6.9) into (5.6.9a) gives

$$V_p\alpha_p + V_0\alpha_0 = V_p\Lambda_p^{-1}U_p^T(U_p\beta_p + U_0\beta_0)$$
$$= V_p\Lambda_p^{-1}\beta_p \qquad (5.6.10)$$

Pre-multiply equation (5.6.10) by V_p^T

$$\alpha_p = \Lambda_p^{-1}\beta_p \qquad (5.6.11)$$

Pre-multiply equation (5.6.10) by V_0^T

$$\alpha_0 = 0 \qquad (5.6.12)$$

Hence the inverse is solved since

$$\hat{x} = V_p\alpha_p \qquad (5.6.13)$$

A solution for x has been achieved when no solution seemed possible.

However, the above discussion gives no indication of the significance of the result that has been obtained. What, it must be asked, has the generalised inverse operator really done? It has minimised two quantities:

1   The sum of the squares of the differences between y and Ax

$$\varepsilon = Ax - y$$
$$|\varepsilon|^2 = |\varepsilon^T\varepsilon| = |(Ax - y)^T(Ax - y)| \qquad (5.6.14)$$

2   The sum of the squares of the parameter values x

$$|x|^2 = |x^T x| \qquad (5.6.15)$$

Equations (5.6.14) and (5.6.15) after transforming from (x, y) to (α, β) and from A to $U\Lambda V^T$ become

$$|\varepsilon|^2 = |\Lambda_p\alpha_p - \beta_p|^2 + |\beta_0|^2 \qquad (5.6.16)$$
$$|x|^2 = |\Lambda_p^{-1}\beta_p|^2 + |\alpha_0|^2 \qquad (5.6.17)$$

Given complete freedom to vary the parameter values α_p and α_0 in equations (5.6.16) and (5.6.17), the minimum values of $|\varepsilon|^2$ and $|x|^2$ would be achieved by setting $\alpha_0 = 0$ and $\Lambda_p\alpha_p = \beta_p$ which is precisely the result obtained in equations (5.6.12) and (5.6.13). Note that when there are no zero eigenvalues (when $p = m$) equation (5.6.12) is not relevant since $\alpha_0 = 0$ by definition.

It is not necessarily desirable that the sum of the squares of the parameter values should be reduced to a minimum. However, if a problem is set up in the manner described in Section 5.5 where x represents a perturbation to a starting model, then the minimisation of $|x|^2$ is saying 'If I am having difficulty finding a solution, I will give you the solution that is nearest to your first guess (starting model) subject to any constraints that the data may impose.'

As a simple example of the application of GMI consider the delay time experiment described in Section 5.4. Travel times and distances are shown in Table 5.1.

Table 5.1.

(i, j)	(t_{ij})	(x_{ij})
1, 1	2.323	6.0
1, 2	2.543	6.708
1, 3	2.857	8.485
2, 1	2.64	7.616
2, 2	2.529	7.0
2, 3	2.553	7.616

The parameter values are:

$$\delta_1 = 0.433$$
$$\delta_2 = 0.346$$
$$\gamma_1 = 0.390$$
$$\gamma_2 = 0.433$$
$$\gamma_3 = 0.303$$
$$V_1 = 4.0$$

The matrix in equation (5.4.4) is decomposed into $U \Lambda V^T$ and the Λ main diagonal elements are shown in Table 5.2.

Table 5.2.

	Singular values	Normalised
1	17.97097067	1.0000E + 00
2	1.73238428	9.6399E − 02
3	1.42082900	7.9062E − 02
4	1.41421356	7.8694E − 02
5	0.15396097	8.5672E − 03
6	0.00000000	7.5362E − 19

The smallest eigenvalue only differs from zero due to computer rounding error. So the problem is ill-conditioned and no inverse can be formed unless the eigenvalue is deleted along with the corresponding vectors within U and V. Referring to equation (5.4.1a) the parameter estimate $\hat{x}$

$$\hat{x} = HAx$$
$$= V_p \Lambda_p^{-1} U_p^T U_p \Lambda_p V_p^T x$$
$$= V_p V_p^T x \qquad (5.6.18)$$

$V_p V_p^T$ is referred to as the resolution matrix. It is not an identity matrix like VV^T because V_p is non-symmetric having lost its sixth column.

The parameter estimates and resolution matrix resulting from an inversion with five non-zero eigenvalues are shown in Table 5.3.

The velocity, V_1, is well resolved and accurately determined. But the values for the delay times are weighted averages of one another. For example:

$$\hat{\delta}_1 = .503 = 0.8 \times .433 - 0.2 \times .346$$
$$+ 0.2(.390 + .433 + .303).$$

Table 5.3.

	Resolution matrix, VV^T					
	δ_1 Col 1	δ_2 Col 2	γ_1 Col 3	γ_2 Col 4	γ_3 Col 5	$\dfrac{1}{V_1}$ Col 6
1	0.80	− 0.20	0.20	0.20	0.20	0.00
2	− 0.20	0.80	0.20	0.20	0.20	0.00
3	0.20	0.20	0.80	− 0.20	− 0.20	0.00
4	0.20	0.20	− 0.20	0.80	− 0.20	0.00
5	0.20	0.20	− 0.20	− 0.20	0.80	0.00
6	0.00	0.00	0.00	0.00	0.00	1.00
Calculated parameters:						
	0.503	0.416	0.301	0.364	0.234	0.25
Actual parameters:						
	0.433	0.346	0.390	0.433	0.303	0.25

It is not surprising that the delay times cannot be determined uniquely since any arbitrary number added to δ_1 and δ_2 and taken away from γ_1, γ_2 and γ_3 in equation (5.4.4) will leave the equations entirely unaffected. To obtain a unique solution for all six parameters, one of the delay times must be specified. This is achieved by adding a seventh constraining equation to equation (5.4.4).

$$\delta_1 = [1\ 0\ 0\ 0\ 0\ 0] \begin{pmatrix} \delta_1 \\ \delta_2 \\ \gamma_1 \\ \gamma_2 \\ \gamma_3 \\ \dfrac{1}{V_1} \end{pmatrix}$$

Then the results of the inversion are shown in Table 5.4.

Table 5.4.

	Singular values	Normalised
1	17.97109279	1.0000E + 00
2	1.89875282	1.0566E − 01
3	1.42190379	7.9122E − 02
4	1.41421356	7.8694E − 02
5	0.63431768	3.5297E − 02
6	0.09896082	5.5067E − 03

	Resolution matrix, VV^T					
	δ_1 Col 1	δ_2 Col 2	γ_1 Col 3	γ_2 Col 4	γ_3 Col 5	$\dfrac{1}{V_1}$ Col 6
1	1.00	0.00	0.00	0.00	0.00	0.00
2	0.00	1.00	0.00	0.00	0.00	0.00
3	0.00	0.00	1.00	0.00	0.00	0.00
4	0.00	0.00	0.00	1.00	0.00	0.00
5	0.00	0.00	0.00	0.00	1.00	0.00
6	0.00	0.00	0.00	0.00	0.00	1.00
Calculated parameters:						
	0.433	0.346	0.390	0.433	0.303	0.25

Note that since no eigenvalues have been deleted, the resolution matrix is inevitably an identity matrix—implying perfect resolution.

It is sometimes useful to have a measure of the relative importance of individual data values in an experiment. Since

$$Hy = HAx = \hat{x}$$
$$\therefore \quad AHy = A\hat{x} = \hat{y}$$
$$\therefore \quad U_p \Lambda_p V_p^T V_p \Lambda_p^{-1} U_p^T y = \hat{y}$$
$$\therefore \quad U_p U_p^T y = \hat{y} \qquad (5.6.19)$$

$U_p U_p^T$ is referred to as the information density matrix.

Table 5.5.

	Information density matrix, UU^T						
	Col 1	Col 2	Col 3	Col 4	Col 5	Col 6	Col 7
1	0.93	0.16	−0.08	0.16	0.08	0.08	0.00
2	0.16	0.67	0.18	−0.16	0.33	−0.18	0.00
3	−0.08	0.18	0.91	0.08	−0.18	0.09	0.00
4	0.07	−0.16	0.08	0.93	0.16	−0.08	0.00
5	−0.16	0.33	−0.18	0.16	0.67	0.18	0.00
6	0.08	−0.18	0.09	−0.08	0.18	0.91	0.00
7	0.00	0.00	0.00	0.00	0.00	0.00	1.00

In the present example the absence of off-diagonal elements in the seventh row or column of the information density matrix is expressing the unique importance of the seventh constraining equation in this inversion.

5.7 Practical problems with GMI

5.7.1 Error in the data

In the preceding sections it has been assumed that the observational data are entirely free from noise. When this highly improbable assumption is discarded, the equation relating data to parameters becomes

$$y + \Delta ye = A(x + \Delta xe) \qquad (5.7.1)$$

where Δye is a vector of noise perturbations to the data and Δxe represents the resulting perturbations to the parameter values. Thus

$$\Delta ye = A\Delta xe \qquad (5.7.2)$$

and following exactly the method described in Section 5.6, an estimate of Δxe is obtained from

$$\Delta \hat{x}e = H_p \Delta ye \qquad (5.7.3)$$

In general, the error in a sequence of n parameters is described by an $(n \times n)$ covariance matrix. It is assumed that an experiment has been run a number of times. Then

$$\langle \Delta xe \, \Delta xe^T \rangle = H_p \langle \Delta ye \, \Delta ye^T \rangle H_p^T \qquad (5.7.4)$$

where the symbol $\langle \rangle$ represents the averaging over all the experimental runs.

If the fluctuations in any data value Δye_i are entirely independent of the fluctuations in any other, Δye_j, then the off-diagonal elements of the covariance matrix $\langle \Delta ye \, \Delta ye^T \rangle$ are all zero. The diagonal elements are the variances (standard deviations squared) of the data.

Assuming that $\langle \Delta ye \, \Delta ye^T \rangle$ is a diagonal matrix in equation (5.7.4) then the main diagonal elements of the matrix $\langle \Delta xe \, \Delta xe^T \rangle$, which are the variances of the parameters, become

$$\text{var}(x_i) = \sum_{j=1}^{n} H_{p_{ij}}^2 \, \text{var}(y_j) \qquad (5.7.5)$$

What equation (5.7.5) means is that parameter resolution is intimately related to parameter variance. Consider a problem in which twelve parameters are to be determined from a set of observational data. An inverse is formed with twelve non-zero eigenvalues. Fig. 5.5a shows the resulting standard deviations of the parameters calculated from equation (5.7.5) and the resolution matrix which is, of course, an identity matrix.

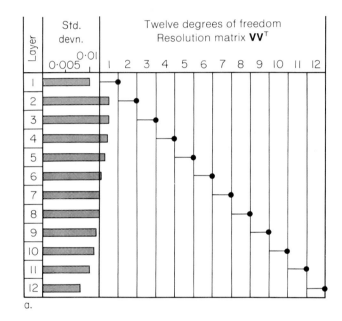

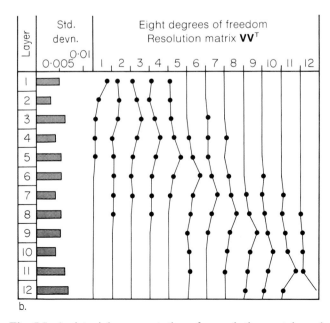

Fig. 5.5. A pictorial representation of a resolution matrix and the error in parameter estimates for inversions with (a) 12 degrees of freedom and (b) 8 degrees of freedom. Note the trade-off between resolution and parameter uncertainty.

The problem is that, whereas in the noise-free case, eigenvalues of the matrix A were either zero or non-zero, in the presence of noise, zero eigenvalues become small but finite and are entirely noise dependent. It is crucial that such terms should be eliminated since the inverse operator H is formed from the inverse of the eigenvalues. The inverse of a small and random number will have a pronounced effect on H, and hence $\hat{x}$. However, there is no precise level below which the eigenvalues are predominantly noise dependent and therefore best deleted (assumed to be effectively zero). The cut-off point is essentially arbitrary.

One is therefore faced with the familiar concept of an uncertainty principle or 'trade-off' between the variance of the parameters and the resolution. Fig. 5.5b shows the standard deviation of the parameters and the resolution matrix resulting from an inversion with eight eigenvalues (the four smallest eigenvalues have been set to zero). The 'bell-like' function defined by any one row of the $V_p V_p^T$ matrix provides a measure of the resolving power of the data. Resolution has therefore been degraded. However, since in equation (5.7.5) var (x_i) is inversely proportional to the eigenvalues of A, the deletion of the four smallest eigenvalues has reduced the variance of the parameter estimates.

Backus and Gilbert introduced the concept of a trade-off curve between variance and spread (defined as the half-width of the bell-like functions in the resolution matrix) (Fig. 5.6). In general, there will be a 'knee' in the curve and they propose that the optimal solution is 'down at the corner' of the trade-off curve. See Gilbert (1972).

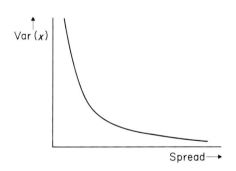

Fig. 5.6. A trade-off curve of parameter resolution against parameter variance.

Assume for the sake of argument, that a model of density versus depth in a laterally homogenous earth has been determined. Consider the apparently simple question 'What is the error in your estimate of density at 2 km depth?' The answer would have to be as follows: 'If you are particularly interested in that precise depth, then between 1900 and 2100 metres, the density is $x \pm$ (some rather large number which may render the value of x to be quite worthless). However, if you are satisfied with an average value of density between 1500 and 2500 metres, then this is tightly constrained by the data to be $x \pm$ (some much smaller number).'

You should also draw attention to any assumptions that were built into the inverse procedure like, for example, the smoothness of the density/depth function. If that does not demoralise the questioner nothing will!

5.7.2 Parameter weighting

The problems with parameter weighting are most conveniently explained with reference to a specific example which requires some explanation. Fig. 5.7 shows a waveform recorded with a 12 ft spacing borehole sonic tool. The first arrival is the refracted compressional wave, travelling with the velocity of the solid medium. The high amplitude and dispersed later arrival is a Stoneley wave. This is an interface wave whose amplitude decays approximately exponentially on both sides of the interface between fluid and formation. The waveform in Fig. 5.7 is comparatively simple because the formation shear velocity is less than the fluid velocity. Under these conditions no refracted shear waves or pseudo-Rayleigh wave can exist. See White (1983) for further information. The practical significance of these data is that, despite the absence of any directly recorded shear wave, it is theoretically possible to obtain estimates of the formation shear velocity from the Stoneley wave arrival.

The phase velocity of the borehole Stoneley wave is frequency dependent and can be related to six parameters: borehole radius, borehole fluid density and velocity, and the compressional velocity, shear velocity and density of the solid medium.

$$c(w) = F(r, \rho_f, V_f, \rho, V_p, V_s) \qquad (5.7.6)$$

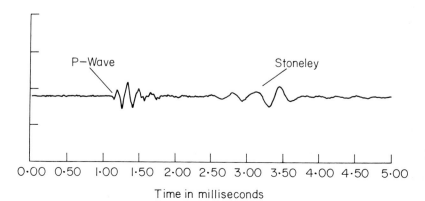

Fig. 5.7. A typical long spacing sonic waveform in a formation where the shear velocity is less than the borehole fluid velocity. Note the first arrival P wave and the Stoneley wave.

Following the method described in Section 5.5, the inverse problem can be linearised as follows:

$$c(w)_{OBS} - \hat{c}(w)_{\substack{\text{starting} \\ \text{model}}} = \frac{\partial \hat{c}}{\partial r} \Delta r + \frac{\partial \hat{c}}{\partial \rho_f} \Delta \rho_f + \frac{\partial \hat{c}}{\partial V_f} \Delta V_f$$
$$+ \frac{\partial \hat{c}}{\partial \rho} \Delta \rho + \frac{\partial \hat{c}}{\partial V_p} \Delta V_p + \frac{\partial \hat{c}}{\partial V_s} \Delta V_s \quad (5.7.7)$$

In the following synthetic example, the model from which $c(w)_{OBS}$ is calculated and the starting model that yields $\hat{c}(w)$ are shown in Table 5.6.

Table 5.6.

	Final model	Starting model	
r	0.11	0.1	(m)
ρ_f	1.65	1.5	(gm/cc)
V_f	1672.0	1520.0	(m/s)
ρ	2.31	2.1	(gm/cc)
V_p	3014.0	2740.0	(m/s)
V_s	1342.0	1220.0	(m/s)

Table 5.7 shows the A matrix of partial derivatives for eight frequency values, and the singular values of the matrix.

The partial derivatives have been calculated numerically (using equation (5.5.5)) with the same percentage change in each parameter. Consequently, the significant differences in the magnitudes of the elements in the columns simply reflects the differing dimensions of the six parameters.

Problems arise when an inversion is carried out with less than six singular values. As discussed in Section 5.6, the effect of such an action is to introduce the additional constraint into the solution of minimising $|\Delta p^T \Delta p|$

$$\Delta p = (\Delta r, \Delta \rho_f, \Delta V_f, \Delta \rho, \Delta V_p, \Delta V_s) \quad (5.7.8)$$

In this case, one would be attempting to minimise

$$(.01)^2 + (.15)^2 + (152)^2 + (.21)^2 + (274)^2 + (122)^2$$

which is most effectively achieved by setting ΔV_f, ΔV_p and ΔV_s to zero, simply because they are the large numbers, and restricting all necessary perturbations to Δr, $\Delta \rho_f$ and $\Delta \rho$. This is what has happened for the inversion with three singular values. Table 5.8 shows the resulting resolution matrix and solution vector Δp.

Table 5.8.

	Δr Col 1	$\Delta \rho_f$ Col 2	ΔV_f Col 3	$\Delta \rho$ Col 4	$\Delta V p$ Col 5	ΔV_s Col 6
1	1.00	0.00	0.00	0.00	0.00	0.00
2	0.00	1.00	0.00	0.00	0.00	0.01
3	0.00	0.00	0.00	0.00	0.00	0.00
4	0.00	0.00	0.00	1.00	0.00	0.02
5	0.00	0.00	0.00	0.00	0.00	0.00
6	0.00	0.01	0.00	0.02	0.00	0.00
$\Delta p =$	-0.21	1.09	-0.001	2.41	0.002	0.05

This is a singularly useless result. The Δp values are wildly off the mark and the resolution matrix is simply expressing the gross disparity in the dimensions of the parameters.

The cure is to weight the parameters so that they are all of a similar size. Thus in general terms:

$$x' = W^{1/2} x$$
$$y = Ax = AW^{-1/2}x' = A'x' \quad (5.7.9)$$
$$A \to AW^{-1/2} \text{ and } \min |x^T x| \to \min |x'^T W^{-1} x'|$$

Having solved for x' using A', x is obtained from:

$$x = W^{-1/2} x' \quad (5.7.10)$$

In the above example $W^{1/2}$ is a diagonal matrix whose

Table 5.7.

	Original Matrix					
	Col 1	Col 2	Col 3	Col 4	Col 5	Col 6
1	-100.0000	-148.0000	0.2237	103.8000	0.0066	0.5148
2	-100.0000	-137.3000	0.1974	96.1900	0.0088	0.5393
3	-100.0000	-128.0000	0.1750	90.4800	0.0117	0.5623
4	-120.0000	-122.7000	0.1566	85.7100	0.0139	0.5787
5	-140.0000	-117.3000	0.1421	82.8600	0.0153	0.5951
6	-160.0000	-114.7000	0.1303	80.0000	0.0168	0.6066
7	-180.0000	-110.7000	0.1211	78.1000	0.0182	0.6164
8	-180.0000	-109.3000	0.1118	76.1900	0.0197	0.6246

	Singular values	Normalised
1	573.79096553	1.0000E + 00
2	95.95224605	1.6723E − 01
3	1.01748769	1.7733E − 03
4	0.08094254	1.4107E − 04
5	0.00202768	3.5338E − 06
6	0.00051736	9.0166E − 07

elements are the inverse of the starting model parameters.

$$W_{ij}^{1/2} = \left(\frac{1}{\hat{r}}, \frac{1}{\hat{\rho}_s}, \frac{1}{\hat{V}_f}, \frac{1}{\hat{\rho}}, \frac{1}{\hat{V}_p}, \frac{1}{\hat{V}_s}\right) \text{ for } i = j$$

$$= 0 \text{ for } i \neq j \qquad (5.7.11)$$

Table 5.9 shows the results of an inversion with three singular values including parameter weighting.

The new parameters are now a reasonable approximation to the final model, particularly V_f and V_s which are the best resolved. This accurately reflects the true physical situation concerning the relative dependence of Stoneley wave velocity on the six parameters.

In order to obtain a better approximation to the true (final) solution, the process would have to be repeated, re-linearising about the new parameter estimates. It must not be forgotten that the inversion is a linear approximation to a non-linear problem and perturbations to the parameter values at each iteration should be kept small (by introducing suitable damping/eigenvalue suppression).

In the example with no parameter weighting, note that $r = -.21$. This means that $r = .11 - .21 = -0.1$ which is physically impossible. One method of ensuring that parameters are never negative is to work with the logarithm of the parameter values:

$$x' = Lnx \qquad (5.7.12)$$

Then even if x' comes out negative, x will always be positive. This is the identical process to the parameter weighting described above. Differentiating equation (5.7.12)

$$\frac{dx_i'}{dx_i} = \frac{1}{x_i} \approx \frac{\Delta x_i'}{\Delta x_i}$$

Therefore

$$\Delta x_i' = \frac{1}{x_i} \cdot \Delta x_i \qquad (5.7.13)$$

which is the same as equations (5.7.10) and (5.7.11).

Also, if $\partial c_j / \partial x_i$ are the elements of the matrix A, then

$$\frac{\partial c_j}{\partial x_i} = \frac{\partial c_j}{\partial x_i'} \cdot \frac{\partial x_i'}{\partial x_i}$$

$$= \frac{\partial c_j}{\partial x_i'} \cdot \frac{1}{x_i}$$

which for all values of i and j is

$$A = A'W^{1/2}$$

as in equation (5.7.9).

Thus correct parameter weighting ensures positivity. (Oristaglio, 1978.)

Table 5.9.

			Original matrix			
	Col 1	Col 2	Col 3	Col 4	Col 5	Col 6
1	−10.0000	−222.0000	340.0240	217.9800	18.0018	628.0560
2	−10.0000	−205.9500	300.0480	201.9990	23.9997	657.9460
3	−10.0000	−192.0000	266.0000	190.0080	32.0032	686.0060
4	−12.0000	−184.0500	238.0320	179.9910	38.0038	706.0140
5	−14.0000	−175.9500	215.9920	174.0060	42.0042	726.0220
6	−16.0000	−172.0500	198.0560	168.0000	46.0046	740.0520
7	−18.0000	−166.0500	184.0720	164.0100	50.0050	752.0080
8	−18.0000	−163.9500	169.9360	159.9990	54.0054	762.0120

	Singular values	Normalised
1	2238.99221470	1.0000E + 00
2	213.12442898	9.5188E − 02
3	5.85057792	2.6130E − 03
4	2.05182119	9.1640E − 04
5	1.71572217	7.6629E − 04
6	0.77464815	3.4598E − 04

	Resolution matrix, VV^T					
	Δr Col 1	$\Delta \rho_f$ Col 2	ΔV_f Col 3	$\Delta \rho$ Col 4	ΔV_p Col 5	ΔV_s Col 6
1	0.35	0.32	0.21	−0.18	−0.21	0.06
2	0.32	0.45	−0.13	−0.32	−0.16	0.00
3	0.21	−0.13	0.82	0.20	−0.21	−0.01
4	−0.18	−0.32	0.20	0.25	0.08	0.03
5	−0.21	−0.16	−0.21	0.08	0.15	0.05
6	0.06	0.00	−0.01	0.03	0.05	0.99
$\Delta p =$	0.007	0.037	152.4	0.001	−107.7	146.9
New parameters:	0.107	1.54	1672.0	2.10	2632.0	1367.0

5.7.3 Data weighting

An analogous procedure to parameter weighting can be implemented if some data values are considered more reliable than others and therefore worthy of a greater influence on the solution. An obvious measure of reliability is the standard deviation of the data. If S is a diagonal matrix with data variance as the main diagonal elements, then the equations

$$y = Ax$$

are appropriately weighted by

$$S^{-1/2}y = S^{-1/2}Ax \qquad (5.7.14)$$

If A is a non-singular, square matrix this equal weighting of both sides of the equation will not have the slightest effect. However, if a solution is sought which minimises $(Ax - y)^T(Ax - y)$ then after weighting, one is minimising $(Ax - y)^T S^{-1}(Ax - y)$. This process is accomplished via the discrete Wiener–Hopf equation which was derived in Section 2.8.2. The derivation of this important equation using matrix notation and including a data weighting term is as follows:

$$E = (Ax - y)^T S^{-1}(Ax - y)$$
$$= (x^T A^T - y^T)(S^{-1}Ax - S^{-1}y)$$
$$= x^T A^T S^{-1}Ax - x^T A^T S^{-1}y - y^T S^{-1}Ax$$
$$\quad + y^T S^{-1}y$$

Minimise E

$$\frac{\partial E}{\partial x^T} = A^T S^{-1}Ax - A^T S^{-1}y = 0$$

Therefore

$$x = [A^T S^{-1}A]^{-1}A^T S^{-1}y \qquad (5.7.15)$$

5.8 Damped least squares

The essence of the discussion in the preceding sections can be stated as follows: if you demand a solution for x from a set of equations $y = Ax$ where no unique solution is possible, fundamentally because the data y do not provide sufficient constraints to define a unique x, then apply some additional constraints on the values that x can take. Generalised matrix inversion automatically applies the constraint that $|x|^2$ should be a minimum. The additional constraints applied by damped least squares inversion are essentially the same but subtly different.

Damping of the solution is achieved by adding equations:

$$\varepsilon x_i = 0 \text{ for } i = 1, \text{ maximum number}$$
$$\text{of parameters}$$

where ε is a constant, known as the damping parameter. Thus

$$y = Ax \qquad (5.8.1)$$

becomes

$$\begin{bmatrix} A \\ \varepsilon I \end{bmatrix} x = \begin{bmatrix} y \\ 0 \end{bmatrix} \qquad (5.8.2)$$

The standard least squares solution to equation (5.8.1) is

$$x = [A^T A]^{-1}A^T y \quad \text{(see Sections 2.8.2 and 5.7.3)}$$

So the solution to equation (5.8.2) is

$$x = \left[[A^T \vdots \varepsilon I]\begin{bmatrix} A \\ \varepsilon I \end{bmatrix}\right]^{-1}[A^T \vdots \varepsilon I]\begin{bmatrix} y \\ 0 \end{bmatrix}$$
$$= [A^T A + \varepsilon^2 I]^{-1}A^T y \qquad (5.8.3)$$

It is also possible to derive equation (5.8.3) from an entirely different starting point. Whereas generalised matrix inversion minimises $|Ax - y|^2$ and $|x|^2$, damped least squares is equivalent to minimising the weighted sum of both

$$|Ax - y|^2 + \varepsilon^2|x|^2$$
$$E = (Ax - y)^T(Ax - y) + \varepsilon^2 x^T x$$
$$= x^T A^T Ax - x^T A^T y - y^T Ax + y^T y$$
$$\quad + y^T y + \varepsilon^2 x^T x \qquad (5.8.4)$$

Minimise E

$$\frac{\partial E}{\partial x^T} = A^T Ax - A^T y + \varepsilon^2 x = 0$$

Therefore

$$[A^T A + \varepsilon^2 I]x = A^T y \qquad (5.8.5)$$

Consider again, as a specific example, the delay time experiment described in Sections 5.4 and 5.6. Table 5.10 shows the augmented matrix equation (5.8.2) with a damping parameter value of 0.05.

Table 5.10.

	Col 1	Col 2	Col 3	Col 4	Col 5	Col 6
			Original matrix			
1	1.0000	0.0000	1.0000	0.0000	0.0000	6.0000
2	1.0000	0.0000	0.0000	1.0000	0.0000	6.7080
3	1.0000	0.0000	0.0000	0.0000	1.0000	8.4850
4	0.0000	1.0000	1.0000	0.000	0.0000	7.6160
5	0.0000	1.0000	0.0000	1.0000	0.0000	7.0000
6	0.0000	1.0000	0.0000	0.0000	1.0000	7.6160
7	0.0500	0.0000	0.0000	0.0000	0.0000	0.0000
8	0.0000	0.0500	0.0000	0.0000	0.0000	0.0000
9	0.0000	0.0000	0.0500	0.0000	0.0000	0.0000
10	0.0000	0.0000	0.0000	0.0500	0.0000	0.0000
11	0.0000	0.0000	0.0000	0.0000	0.0500	0.0000
12	0.0000	0.0000	0.0000	0.0000	0.0000	0.0500

	Singular values	Normalised
1	17.97104022	1.0000E + 00
2	1.73310568	9.6439E − 02
3	1.42170850	7.9111E − 02
4	1.41509717	7.8743E − 02
5	0.16187644	9.0076E − 03
6	0.05000000	2.7823E − 03

Compare the eigenvalues of this matrix (Table 5.10) with those of the original A matrix (Table 5.2, Section 5.6). The relationship is:

$$\lambda_i(\text{NEW}) = (\lambda_i^2(\text{OLD}) + \varepsilon^2)^{1/2} \qquad (5.8.6)$$

The result is that when $\varepsilon \ll \lambda_i$, $\lambda_i(\text{NEW}) \approx \lambda_i(\text{OLD})$.

However, the inclusion of damping ensures that no eigenvalue is less than ε and hence the existence of a least squares inverse is guaranteed. The choice of damping parameter value depends on the noise level in the data. As signal/noise gets worse, the noise level in the eigenvalues increases. As a rule of thumb, ε should be chosen to be equal to the largest eigenvalue that is believed to be predominantly noise dependent.

The equivalent operation using generalised matrix inversion would involve chopping off eigenvalues that are smaller than some chosen value. In the presence of noise, this cut-off value is inevitably somewhat arbitrary. Therefore the gradational weighting of the eigenvalues that results from damped least squares is an intuitively more satisfactory approach and in practice is easier to implement.

In Section 5.4, an iterative procedure for solving non-linear inverse problems was described. Convergence can be a problem if the $(n + 1)$ dimensional field of $\sum_i(y_i - \hat{y}_i)^2$ versus n parameters is a complicated hypersurface of local maxima and minima. It is usually necessary to control the step length Δx at each iteration in some systematic manner, so that initially it is large but reduces as a stable minimum as $\sum_i(y_i - \hat{y}_i)^2$ is approached. This is most conveniently achieved using damped least squares. The size of the damping parameter controls the step length Δx. As ε increases, Δx decreases, until for very heavy damping, no perturbation to a particular starting model will occur at all.

Finally, referring back to Section 2.8.4, it should be appreciated that the introduction of white light in the construction of deconvolution filters is a direct application of damped least squares inversion. The terms in equation (5.8.5) represent the following operations:

$A^T Ax$: The auto-correlation function of the input convolved with the filter x.

$\varepsilon^2 Ix$: An impulse function of amplitude ε^2 convolved with the filter x.

$A^T y$: The input to the filter convolved with the output from the filter.

The physical significance of damping is most clearly displayed by transforming equation (5.8.5) into the frequency domain. Hence

$$[F_i^*(w)F_i(w) + \varepsilon^2]X(w) = F_i^*(w)F_0(w) \qquad (5.8.7)$$

where $F_i(w) =$ the Fourier transform of the input waveform to the filter and $*$ represents the complex conjugate; $F_0(w) =$ the Fourier transform of the output waveform from the filter; $X(w) =$ the Fourier transform of the impulse response of the filter; ε^2 is a constant for all values of w. Hence,

$$x = F.T.^{-1}\left\{\frac{F_i^*(w)F_0(w)}{F_i^*(w)F_i(w) + \varepsilon^2}\right\} \qquad (5.8.8)$$

If $\varepsilon = 0$, large values of the bracketed term in equation (5.8.8) will result from small values of $F_i(w)$. Normally, the smaller the values of $F_i(w)$, the worse will be the signal/noise ratio and a situation can easily arise in which the largest values of $X(w)$ are more dependent on the noise than the signal. This undesirable amplification of noise can be prevented by the judicious choice of the damping parameter ε. The presence of the term ε^2 in the denominator of equation (5.8.8) ensures that very small values of $F_i(w)$ are never permitted to dominate the solution.

White light can be criticised on the grounds that its usage affects the gain at all frequencies, even though this is not desirable at those frequencies where F_i is well represented. In some unpublished work done in 1976, Hatton and C. Hewlett showed how light could be added to equation (5.8.5) in such a way as to affect only desired areas of the frequency domain. They called this technique 'coloured light' stabilization. The coloured light equivalent to equation (5.8.5) is

$$[A^T A + \lambda C]x = A^T y \qquad (5.8.9)$$

where λ is scalar and C is a colour matrix. C has Toeplitz structure, i.e.:

$$\begin{bmatrix} C_0 & C_1 & C_2 \dots\dots\dots\dots\dots & C_N \\ C_1 & C_0 & C_1 & C_2 \dots\dots\dots & C_{N-1} \\ C_2 & C_1 & C_0 & C_1 & C_2\dots\dots & C_{N-2} \\ \dots\dots\dots\dots\dots\dots\dots\dots\dots\dots\dots\dots \\ \dots\dots\dots\dots\dots\dots\dots\dots\dots\dots\dots\dots \\ \dots\dots\dots\dots\dots\dots\dots\dots\dots\dots\dots\dots \\ C_N & C_{N-1}\dots\dots\dots\dots\dots & C_0 \end{bmatrix} = C$$

The colour coefficients C_i may be conveniently determined in the knowledge that their Fourier transform has some desired shape for filling up 'holes' or regions of poor signal-to-noise ratio in the frequency domain, for example, the notch due to the presence of the 'ghost' as was discussed in Section 2.3. After specifying this shape in the frequency domain, the colour coefficients can be determined by a simple inverse Fourier transform. Hatton and Hewlett used Gaussian humps, modifying the heights and widths as appropriate. Such Gaussian humps have an analytic inverse transform making the colour coefficients particularly easy to evaluate.

Their results suggested that in excellent signal-to-noise conditions, the usage of colour could be beneficial. On conventional seismic data however, there was no observable difference, the noise already present in the data serving to distort the spectrum sufficiently to render the technique irrelevant.

5.9 VSP processing

The methods and techniques of imaging the earth which are described in Chapters 3 and 4 are all based upon horizontal source–receiver array configurations. If the images that are obtained instil sufficient confidence and hope into the hearts of management, a borehole will be drilled. Thereafter an additional suite of data are potentially available to the seismic processor which can lead to significant changes to the images upon which the original drilling decision was based.

It was emphasised in Chapter 4 that by far the most important input parameter to any seismic migration algorithm is velocity. At the very least, a borehole can provide some additional velocity estimates. To be more precise, the borehole provides a time to depth transform which in principle can be deduced by integrating the transit times recorded by a downhole sonic tool as it moves through the rock section. In practice, a check shot survey is also required to calibrate the sonic data. A shot is fired close to the well head and is recorded by a clamped geophone at various depths down the hole. Travel times measured from the check shot survey can differ from the integrated sonic times for a variety of reasons: the very existence of the borehole will inevitably have perturbed the nature of the rocks within the immediate vicinity of the hole to some degree; the frequencies of the sonic data and check shot data differ by a factor of a thousand and the small degree of dispersion in seismic body waves is detectable over this frequency range. In addition, the sonic tool is rarely run right up to the surface. So a check shot time is required to tie the top of the sonic run to a zero depth level. Corrections are applied to the sonic data so that integrated sonic times and check shot times are mutually consistent over all depth intervals.

The check shot survey is a very simple experiment essentially only involving the measurement of first arrival times of the down-going waves. However, a great deal more information than one travel time can be gleaned from the recording from a buried geophone. It is intuitively obvious that there must be some advantage in filling a borehole with geophones if one is particularly interested in the earth structure within the immediate vicinity of the hole. Experiments involving various configurations of surface and buried sources

and receivers are referred to under the general heading, vertical seismic profiling (VSP).

The objectives of VSP experiments fall into two broad categories which are in no sense mutually exclusive:

1 Surface seismic data processing can be influenced by knowledge gain from VSP data of how the wave-field changes as it propagates through the earth. For example: deconvolution filters can be designed on the basis of VSP wavelets; and multiples can be positively identified from the VSP data.

2 The resolution of a seismic image within the immediate vicinity of a borehole can, in principle, be improved by moving the sources or receivers as close as possible to the region of interest; in other words, down the hole.

Pursuit of the second objective frequently requires surveys in which a large number of shots are fired at differing offsets and azimuths from the well head, and geophones are positioned at as many levels in the borehole as time and cost permit. Such surveys are referred to as offset VSP.

Most VSP processing is not fundamentally different from the methods described in Chapters 3 and 4. CDP stacking and migration can be adapted to a two-dimensional source–receiver geometry. However, in one important respect, VSP data are different from surface data. The wave field approaches the receivers from above and below. Special procedures are required to separate the upgoing from the downgoing waves and one such procedure, based on the work of Seeman and Horowicz (1983) with minor modifications by Stainsby (1984) is described below. For a lead into the extensive literature on VSP see Balch and Lee (1984) and Cassell (1984).

5.9.1 Separation of upgoing and downgoing waves

Fig. 5.8 shows the results of a hypothetical VSP survey. Note that a primary reflection (upgoing wave) always terminates on the first arrival (downgoing wave) branch.

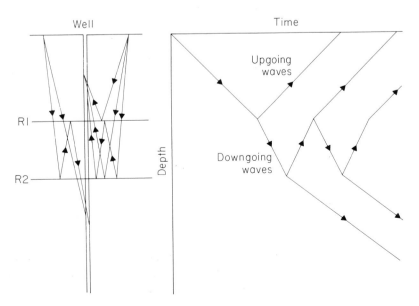

Fig. 5.8. A synthetic VSP dataset.

All multiple branches, whether downgoing or upgoing, always terminate at later times than the first arrival branch. Hence, multiple energy can be positively identified.

Note also that the apparent velocities of the upgoing and downgoing energy are of opposite sign. So, in principle, separation should be possible by using f–k filtering as described in Section 2.10. The main objection to this approach is that f–k filtering is based upon the fast Fourier transform algorithm which assumes regular sampling in time and space. Geophone positions in a borehole will not necessarily be at constant depth intervals. Positions are often adjusted to avoid 'washed-out' regions of the hole, which could be the cause of poor coupling between the geophone and the rock.

An alternative approach is to consider the separation of upgoing and downgoing waves as a linear inverse problem and use one of the methods described in the previous sections of this chapter.

Consider a very simple example. Three geophones are positioned in a vertical borehole above a horizontal reflector. Assume that the wavelet shape does not change appreciably as it propagates from one geophone position to the next. The signals recorded at each geophone can be expressed as follows:

$$S_0(t) = u_0(t) + d_0(t) + n_0(t)$$

$$S_1(t) = u_0(t - \Delta t) + d_0(t + 2\Delta t) + n_1(t)$$

$$S_2(t) = u_0(t - 2\Delta t) + d_0(t + 2\Delta t) + n_2(t)$$

$$\Delta x_1 = V_1 \Delta t$$
$$\Delta x_2 = V_2 \Delta$$

(5.9.1)

u_0 and d_0 are the upgoing and downgoing waves respectively at level 0. n_0, n_1 and n_2 are the components of noise at the three levels. The time delay between levels 0 and 1 is the same as the time delay between levels 1 and 2.

In the frequency domain, equations (5.9.1) become:

$$S_0(f) = U_0(f) + D_0(f) + N_0(f)$$

$$S_1(f) = e^{-iw\Delta t}U_0(f) + e^{iw\Delta t}D_0(f) + N_1(f)$$ (5.9.2)

$$S\ (f) = e^{-iw2\Delta t}U_0(f) + e^{iw2\Delta t}D_0(f) + N_2(f)$$

For any frequency

$$\begin{pmatrix} S_0 \\ S_1 \\ S_2 \end{pmatrix} = \begin{bmatrix} 1 & 1 \\ e^{-iw\Delta t} & e^{iw\Delta t} \\ e^{-iw2\Delta t} & e^{iw2\Delta t} \end{bmatrix} \begin{pmatrix} U_0 \\ D_0 \end{pmatrix} + \begin{pmatrix} N_0 \\ N_1 \\ N_2 \end{pmatrix}$$ (5.9.3)

Writing equation (5.9.3) as

$$y = Ax + n$$ (5.9.4)

the least squares solution for the upgoing and downgoing waves at level 0 is

$$x = \begin{pmatrix} U_0 \\ D_0 \end{pmatrix} = [\tilde{A}^T A]^{-1} \tilde{A}^T y$$ (5.9.5)

where $\tilde{A}$ represents the complex conjugate of A.

The full procedure in the general case is as follows:

1 Calculate the Δt values between the geophone positions. It is not necessary to assume that the wavefront is travelling parallel to the borehole. By replacing the velocities by apparent velocities, slanting boreholes and dipping reflecting boundaries can be handled.
2 Form the A matrix with the results from 1.
3 Fourier transform the data recorded at each level and form y for each frequency component.
4 Solve equation (5.9.5) for all frequency components.
5 Inverse Fourier transform the result to obtain time domain estimates of the upgoing and downgoing wave fields at any desired reference level.

Unfortunately, situations do arise in which $[\tilde{A}^T A]$ is singular. Consider again the simple example above. In that case,

$$[\tilde{A}^T A]^{-1} = \frac{1}{Det[\tilde{A}^T A]}$$
$$\times \begin{bmatrix} 3 & 1 + e^{iw2\Delta t} + e^{iw4\Delta t} \\ 1 + e^{-iw2\Delta t} + e^{-iw4\Delta t} & 3 \end{bmatrix}$$

(5.9.6)

and $Det[\tilde{A}^T A] = (6 - 4 \cos w2\Delta t - 2 \cos w4\Delta t)$

Note that

$$Det[\tilde{A}^T A] = 0 \quad \text{when} \quad f = \frac{k}{2\Delta t}, \quad k = 0, 1, 2 \dots$$

At frequencies that are very close to these critical values, $Det[\tilde{A}^T A]$ is finite but very small. So the data vector y is divided by a very small number which is fine for perfect noise-free data and disastrous if that particular frequency component of the real data is noise dominated.

This type of problem has been encountered in other examples earlier in the chapter and a convenient solution is to use damped least squares (Stainsby, 1984). Equation (5.9.5) becomes

$$\begin{pmatrix} U_0 \\ D_0 \end{pmatrix} = [\tilde{A}^T A + \gamma I]^{-1} \tilde{A}^T y$$ (5.9.7)

where γ is the damping parameter.
In our simple example,

$$Det[\tilde{A}^T A + \gamma I]$$
$$= (6 + 6\gamma + \gamma^2 - 4 \cos w2\Delta t - 2 \cos w4\Delta t)$$

(5.9.8)

The presence of γ ensures that an inverse always exists, and stabilises the inverse against the possibility of obtaining a noise dominated solution.

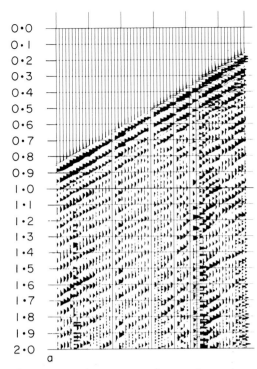
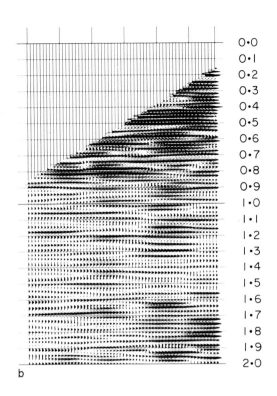

Fig. 5.9. VSP data processed to separate the upgoing and downgoing wave fields.

One further practical detail must be mentioned. It is virtually always the case that downgoing wave amplitudes will be very much greater than upgoing wave amplitudes. So if one is going to use damped least squares, which minimises

$$|Ax - y|^2 + \gamma |x|^2,$$

the same problems will be encountered as were described in Section 5.7.2. Parameter weighting must be included. The appropriate weighting matrix is

$$W = \begin{bmatrix} 1 & 0 \\ 0 & G \end{bmatrix}$$

where G is the amplitude ratio of the downgoing to upgoing waves.

The data shown in Fig. 5.9 are from a land well VSP. Note that the data have been shifted so that the first arrivals correspond to two-way time. Consequently the up going P wave energy should align approximately horizontally. Fig. 5.9a shows the strong downgoing P wave energy and also some weaker tube wave and mode converted shear wave energy. Fig. 5.9b shows the upgoing wave energy that has resulted from processing the data in Fig. 5.9a. Signature deconvolution based on the extracted downgoing wavelet has also been implemented. Note that the data in Fig. 5.9b show evidence of dipping reflectors and possibly faulting close to the well.

5.10 Residual statics

Another important application of damped least squares inversion is in the analysis of residual statics (Wiggins *et al.*, 1976). This concept was introduced and defined in Section 3.7.1. Essentially, an attempt is made to correct for time shifts in the reflection seismic data that are due

solely to near surface lateral variations in the velocity structure. Refer to Fig. 3.97 and consider one ray path from source to buried reflector to receiver. The total travel time from shot position i to receiver position j, is T_{ij},

$$T_{ij} = S_i + R_j + G_k + M_k X_{ij}^2 \qquad (5.10.1)$$

where

S_i = the travel time from the source to the datum plane at position i

R_j = the travel time from the datum plane to the receiver at position j

G_k = the normal incidence two-way travel time from the datum plane to a subsurface reflector at the kth CDP position.

The term $M_k X_{ij}^2$ represents the residual normal moveout time for a source–receiver separation of X_{ij}. For the source–receiver locations illustrated in Fig. 5.10 the travel time equations can be expressed as the matrix equation,

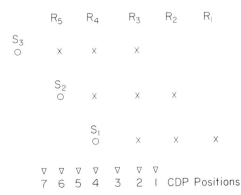

Fig. 5.10. A simple configuration of shot positions, receiver positions and CDP positions, relevant to the discussion of statics analysis.

$$
\begin{bmatrix} T_{11} \\ T_{12} \\ T_{13} \\ T_{22} \\ T_{23} \\ T_{24} \\ T_{33} \\ T_{34} \\ T_{35} \end{bmatrix} =
\begin{bmatrix}
1\,0\,0 & 1\,0\,0\,0\,0 & 1\,0\,0\,0\,0\,0\,0 & 9\,0\,0\,0\,0\,0\,0 \\
1\,0\,0 & 0\,1\,0\,0\,0 & 0\,1\,0\,0\,0\,0\,0 & 0\,4\,0\,0\,0\,0\,0 \\
1\,0\,0 & 0\,0\,1\,0\,0 & 0\,0\,1\,0\,0\,0\,0 & 0\,0\,1\,0\,0\,0\,0 \\
0\,1\,0 & 0\,1\,0\,0\,0 & 0\,0\,1\,0\,0\,0\,0 & 0\,0\,9\,0\,0\,0\,0 \\
0\,1\,0 & 0\,0\,1\,0\,0 & 0\,0\,0\,1\,0\,0\,0 & 0\,0\,0\,4\,0\,0\,0 \\
0\,1\,0 & 0\,0\,0\,1\,0 & 0\,0\,0\,0\,1\,0\,0 & 0\,0\,0\,0\,1\,0\,0 \\
0\,0\,1 & 0\,0\,1\,0\,0 & 0\,0\,0\,0\,1\,0\,0 & 0\,0\,0\,0\,9\,0\,0 \\
0\,0\,1 & 0\,0\,0\,1\,0 & 0\,0\,0\,0\,0\,1\,0 & 0\,0\,0\,0\,0\,4\,0 \\
0\,0\,1 & 0\,0\,0\,0\,1 & 0\,0\,0\,0\,0\,0\,1 & 0\,0\,0\,0\,0\,0\,1
\end{bmatrix} \boldsymbol{p} \qquad (5.10.2)
$$

where

$$\boldsymbol{p}^T = (S_1 \to S_3, \quad R_1 \to R_5, \quad G_1 \to G_7, \quad M_1 \to M_7)$$

$$\boldsymbol{t} = A\boldsymbol{p} \qquad (5.10.3)$$

Normally, S and R will represent the residual static time shift that remains after a first attempt has been made to correct the travel times to the datum level on the basis of field observations (e.g. surveying of the topography and estimates of the near surface velocity from up-hole shot times). Also, a normal moveout correction will have been applied. So M represents a residual normal moveout coefficient. It is common practice to average the M coefficients over some distance window along the profile since they are unlikely to vary very greatly from one CDP position to the next. So M_i to M_{i+n} is reduced to M_j (average) and columns i to $i + n$ of the matrix A are summed together to produce one column. Note the surface consistency assumption inherent in equation (5.10.2). S_i and R_j are assumed to be independent of the offset and hence the take-off angle of the rays.

The solution of equation (5.10.3) is an undetermined inverse problem. Although the number of observations will usually exceed the number of unknowns, a singular value decomposition of the matrix A will include fewer non-zero eigenvalues than the number of unknowns. Referring to Section 5.8, a solution may be obtained to this problem by damped least squares, in which the damping parameter controls the extent of the perturbations to the parameter values. Equation (5.8.5) re-expressed with appropriate symbols becomes:

$$[A^T A + e^2 I]\boldsymbol{p} = A^T \boldsymbol{t} \qquad (5.10.4)$$

In practice, the matrix $[A^T A + e^2 I]$ is never inverted directly. Given the number of source and receiver positions in a normal reflection survey, no existing digital computer could complete the task in anything like a reasonable time. (See Section 5.12 for further discussion on the inversion of large matrices.)

An iterative solution like Gauss-Seidel (see Noble and Daniel, 1977) normally produces an acceptable degree of convergence after about four or five iterations, and the computer run time, compared with direct inversion is drastically reduced.

5.11 Linear programming

Linear programming is the name given to problems that can be formulated as follows:

Maximise the function $z = c^T x$

subject to the condition that $Ax \leqslant b, \; x > 0$

$$(5.11.1)$$

Fig. 5.11 shows an example when $x = (x_1, x_2)$.

The problem is solved by moving the line defined by $c^T x$ as far as possible in the direction of the arrow without leaving the region bounded by the linear constraints (shown shaded in Fig. 5.11 and referred to as the feasible region).

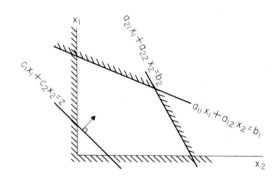

Fig. 5.11.

Most mainframe computer systems will contain at least one linear programming package within their subroutine library, since a very wide variety of problems can be coaxed into this general form.

For the purposes of this discussion, the linear programming subroutine will be considered as a 'black box' (see Vajda (1970) for algorithmic details). With regard to the input parameters, versions vary as to the degree of flexibility allowed in the definition of the linear constraints. The definition at the beginning of this section appears rather restrictive. However, this is really not so, since:

1 maximise $c^T x \equiv$ minimise $- c^T x$
2 $Ax \geqslant b \equiv - Ax \leqslant - b$
3 $x > 0 \equiv x = x^+ - x^-$ where x^+ and x^- are > 0 and x can have either sign.

Two geophysical applications of this readily available software will now be considered.

5.11.1 Extremal inversion

Two extremal inversion methods have already been described in this chapter, namely Monte Carlo and Hedghog inversion. The objective is to determine the maximum permissible range of model parameter values rather than any one acceptable model. Note that the upper and lower bounds of parameter space are not themselves models that satisfy the data.

Consider again the problem of determining borehole parameters from Stoneley wave phase velocities, which was described in Section 5.7.2. Linear programming can be used to determine the maximum and minimum values of the six parameters (Johnson, 1972) in the following manner:

$$c(w) - \hat{c}(w) = \sum_{i=1}^{6} \frac{\partial \hat{c}}{\partial p_i} (p_i - \hat{p}_i) \qquad (5.11.2)$$

where $p = (r, \rho_f, V_f, \rho, V_p, V_s)$ and all symbols have been defined in Section 5.7.2. Rearranging (5.11.2)

$$\sum_{i=1}^{6} \frac{\partial \hat{c}}{\partial p_i} p_i = c(w) - \hat{c}(w) + \sum_{1}^{6} \frac{\partial \hat{c}}{\partial p_i} \hat{p}_i \qquad (5.11.3)$$

If $\sigma(w)$ represents some confidence limit on the observed value $c(w)$, then

$$\sum_{i=1}^{6} \frac{\partial c}{\partial p_i} p_i \geqslant c(w) - \sigma(w) - \hat{c}(w) + \sum_{i=1}^{6} \frac{\partial \hat{c}}{\partial p_i} \hat{p}_i$$

$$\sum_{i=1}^{6} \frac{\partial \hat{c}}{\partial p_i} p_i \leqslant c(w) + \sigma(w) - \hat{c}(w) + \sum_{i=1}^{6} \frac{\partial \hat{c}}{\partial p_i} \hat{p}_i$$

$$(5.11.4)$$

The maximum value of V_s, for example, is then obtained by maximising $c^T x$ where

$$c = (0, 0, 0, 0, 0, 1)$$

$$x = p$$

and equation (5.11.4) are the linear constraints.

Table 5.11 shows the results of such a linear programming extremal inversion of the Stoneley wave from Section 5.7.2. Confidence limits on the phase velocities were set at ± 1 m/sec.

Table 5.11.

Parameter	Lower bound	Actual value	Upper bound
r	0	0.11	0.244
ρ_f	0.063	1.65	2.91
V_f	164.9	1672.0	2992.0
ρ	0	2.31	6.197
V_p	0	3014.0	7583.0
V_s	1173.0	1342.0	1535.0

Clearly, shear velocity is the only parameter that can be reliably estimated from these data, a conclusion that is consistent with the resolution matrix in Table 5.9, Section 5.7.2.

One word of caution is necessary. The foregoing inversion is based upon a linear approximation to a non-linear equation. As $(p_i - \hat{p}_i)$ gets large, the approx-imation gets worse. Consequently the extremal regions of parameter space are subject to the maximum linear-isation error. However, this is only a serious problem when σ is considerably larger than $(c(w) - \hat{c}(w))$.

5.11.2 L_1 norm inversion

The most frequently used criterion for the goodness of fit of calculated to observed data is the sum of the square of the differences (the L_2 norm). One alternative is to use the sum of the absolute values of the differences which is known as the L_1 norm. Minimise

$$\sum_{i=1}^{n} |m - x_i| \qquad (5.11.5)$$

The main advantage of using this criterion is that it is less sensitive than the L_2 norm to the presence of a few highly erratic data points in an otherwise relatively noise free dataset. (Claerbout and Muir, 1973.)

L_1 norm inversion is most conveniently accomplished by using a linear programming routine.
If

$$y = Ax + \varepsilon$$

let

$$x = x^+ - x^-$$

$$\varepsilon = \varepsilon^+ - \varepsilon^-$$

and

$$[A \vdots -A \vdots I \vdots -I]\begin{bmatrix} x^+ \\ x^- \\ \varepsilon^+ \\ \varepsilon^- \end{bmatrix} = d \qquad \begin{pmatrix} x^+ \\ x^- \\ \varepsilon^+ \\ \varepsilon^- \end{pmatrix} \geqslant 0$$

$$(5.11.6)$$

Then one simply minimises z

$$z = (0, 0, 1, 1) \begin{pmatrix} x^+ \\ x^- \\ \varepsilon^+ \\ \varepsilon^- \end{pmatrix} \qquad (5.11.7)$$

In practice, if your system has an LP package, there is a good chance it will also have an L_1 norm routine which sets up the above matrices and vectors for you.

5.12 Tomography

The various procedures described in the preceding sections of this chapter have all been potentially applicable to any geophysical inverse problem. This section is specifically concerned with inverse problems in which a line integral relationship exists between the observed data and the earth parameters. For example, the seismic velocity and attenuating properties of the earth can be related to the observed travel time (t_k) and amplitude (a_k) of a seismic wave by a line integral along a ray path (R_k).

Seismic tomography is the inversion of these line integral relationships (see Fig. 5.12) to obtain estimates of

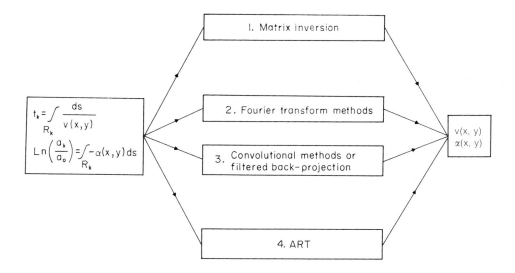

Fig. 5.12.

the velocity field, $v(x, y)$, or the attenuation field, $\alpha(x, y)$, within some region of space through which the rays have passed. Four distinct inverse procedures are considered.

5.12.1 Matrix inversion

The methods of generalised matrix inversion, damped least squares or linear programming are all readily applied to this problem. The region of interest is divided into a grid of cells, and $v(x, y)$ or $\alpha(x, y)$ are assumed to be constant over the area covered by any one cell. Taking the travel time equation as an example, the line integral may be approximated as:

$$t_k = \sum_j \frac{\Delta s_j}{v_j} \qquad (5.12.1)$$

where

Δs_j = the distance travelled by the ray through cell j
v_j = the seismic velocity within cell j

and the summation is taken over the cells actually intersected by the kth ray path. In practice it is advisable, for the reasons given in Section 5.6, to set up the equations so that one is solving for the perturbation of the velocity values from some starting model. Thus

$$\Delta t_k = t_k^{OBS} - t_k^{CALC} = \sum_j \Delta s_j \cdot \Delta p_j \qquad (5.12.2)$$

where $\Delta p_j = 1/v_j - 1/\hat{v}_j$ is the perturbation of the slowness (reciprocal of velocity) within the jth cell. In terms of the equation $y = Ax$, A is a (k max $\times j$ max) matrix of Δs values where k max is the total number of rays crossing the region of interest and j max is the total number of cells. A is a relatively sparse matrix (contains a significant number of zero elements) since any one ray normally only intersects a small fraction of the cells within the region under investigation.

In order to set up the A matrix it is necessary to know the path that the rays take from source to receiver. This will depend upon the velocity field, and so it would appear that the solution to the problem is

required to obtain the solution! In practice, the procedure described in Section 5.5 can be implemented. Rays are traced through an initial (guessed) model and the matrix A is constructed. The equation $y = Ax$ is solved using generalised matrix inversion, damped least squares or linear programming inversion. Perturbations to the velocity, contained within x, are kept small by suitable damping within the inversion and the deviations to the ray paths that would result are also assumed small and ignored. Rays are then traced through the new modified velocity model which results in a new A matrix. This process may be repeated as many times as is required in order to minimise the difference between observed and calculated travel times, to some specified accuracy.

The main requirement for a successful tomographic inversion is that the angular coverage of the rays should be as wide as possible. In most geophysical applications, source–receiver positioning is severely restricted. This will inevitably give rise to blurring and ambiguity in the image that is achieved. Fig. 5.13 shows a rather extreme example. 140 rays have been traced through the simple velocity model shown in Fig. 5.13a from 10 shot positions to 14 receiver positions. The region of interest is 8 cells wide and 12 cells long. Hence, A is a 140×96 matrix. A singular value decomposition of the matrix A revealed that only 70 out of the 96 singular values were non-zero (or effectively non-zero). This is not surprising since some of the cells in the bottom right-hand corner of the region are not intersected by any rays at all. Inversion with 70 degrees of freedom ($p = 70$ in the nomenclature of Section 5.6) results in the image in Fig. 5.13b. Fig. 5.13c is the image that results from inverting the travel times from only one shot. This was achieved using damped least squares. Initially A is a (14×96) matrix, which is effectively augmented to (110×96) by the constraint equations $\varepsilon x_i = 0$ (see Section 5.8). The smearing of the image in the direction of the ray paths is an inevitable consequence of the very limited directional coverage in this example.

The major disadvantage of matrix inversion applied to tomography is the very large computational effort

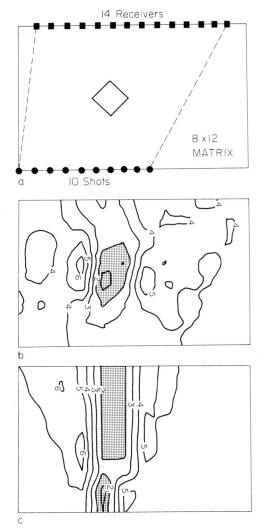

Fig. 5.13. An example of a tomogram of a velocity field obtained using generalised matrix inversion. (a) The velocity field consists of a square anomaly within an otherwise homogeneous region. The velocity contrast is 10 per cent. 140 rays are traced through the region from 10 shots to 14 receivers. The region is divided, for inversion purposes, into 96 blocks. (b) The resultant inversion. (c) The resultant inversion when only the 14 rays from one shot are available. Note the inevitable lack of resolution in the shot to receiver direction.

that is usually required. The number of operations involved in inverting a matrix is of the order of n^3, where n is the number of cells within the region of interest. Consequently (10×10) cells takes 10^6 time units and (100×100) cells takes 10^{12} time units. So at 1 μsec per operation, the (100×100) cells will take 277 hours to invert. (200×200) cells would take two years! (100×100) cells is not unrealistic in many applications, so it is vital that other more rapid computational procedures should be available. The next two methods provide the speed but at the expense of restrictions on the source–receiver configuration of the experiment, and the assumed ray paths.

5.12.2 Fourier transform methods

Fourier transform methods are based upon the projection slice theorem which states that: the one-dimensional Fourier transform of a projection at an angle θ is a slice at the same angle of the two-dimensional Fourier transform of the original object (Mersereau and Oppenheim, 1974).

Consider Fig. 5.14. Fig. 5.14a shows the velocity field which is a square of high velocity within an otherwise homogeneous velocity field. (In fact the edges of the square have been slightly smoothed for computational reasons which should not distract the reader from the main argument.) The projection of this velocity field is the travel times of parallel rays that pass across the field at a constant angle. When the rays are parallel to the x or y axes, the travel time anomaly is a box. However, when the ray paths are at 45 degrees to the x or y axes, the projected travel time anomaly will be a triangle. Fig. 5.14b shows the two-dimensional Fourier transform of Fig. 5.14a. According to the projection slice theorem, a plane through Fig. 5.14b will be the one-dimensional Fourier transform of a projection through Fig. 5.14a at the same angle. Note that slices along lines A and B in Fig. 5.14b are the Fourier transforms of a box and a triangle respectively.

Suppose that parallel straight rays are passed through the region of interest in the x_1 direction (see Fig. 5.15). The projection of the rays is

$$p_{x_1}(x_2) = \int_A^B f(x_1, x_2)\, dx_1 \tag{5.12.3}$$

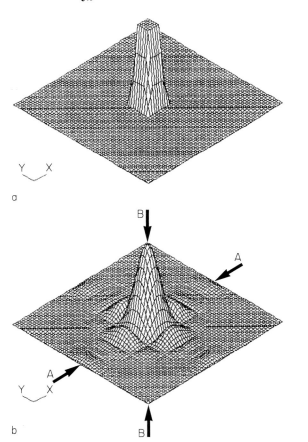

Fig. 5.14. A pictorial representation of the projection slice theorem (see text for details). (b) is the two-dimensional Fourier transform of (a). A plane through (b), plane A–A or B–B for example, is the one-dimensional Fourier transform of a projection through (a) at the same angle.

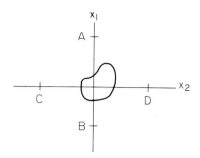

Fig. 5.15.

p represents travel times or amplitudes depending on the application, and $f(x_1, x_2)$ is the property of the medium to be determined. The Fourier transform of $p_{x_1}(x_2)$ is

$$P_{x_1}(k_2) = \int_C^D p_{x_1}(x_2)e^{-ik_2x_2}\,dx_2 \qquad (5.12.4)$$

But the two-dimensional Fourier transform of $f(x_1, x_2)$ is

$$F(k_1, k_2) = \int_C^D \int_A^B f(x_1, x_2)e^{-i(k_1x_1 + k_2x_2)}\,dx_1\,dx_2 \qquad (5.12.5)$$

Substitute (5.12.3) into (5.12.4) and compare the result with (5.12.5) and one obtains

$$P_{x_1}(k_2) = F(k_1, k_2)\bigg|_{k_1} \qquad (5.12.6)$$

The above proof can be generalised for any arbitrary projection angle θ (see Fig. 5.16) (Mersereau and Oppenheim, 1974).

In principle, this permits a very rapid method of back-projection. One Fourier transforms all the projections at the different angles in order to construct a two-dimensional Fourier transform field like Fig. 5.14b.

and the final image can be seriously degraded as a result. In addition, interpolation is computationally very expensive. For this reason, an equivalent convolutional approach turns out to be just as efficient.

5.12.3 Convolutional methods or filtered back-projection

The most frequent use of this method in geophysics is not for imaging at all but in slant stacking or τ-p processing of seismic data (Robinson, 1982).

If the field (x, t) in Fig. 5.17 consists of CDP gathered data, then the summation of amplitudes along a line with intercept τ and gradient p represents a slant stack. The summation can then be plotted as a point in τ, p space.

In precisely the same way, suppose t were another space coordinate, depth. Shots are fired at various depths t in a borehole at $x = 0$, and seismic arrivals are recorded in a vertical borehole at $x_{\max}$. The (x, t) field now represents the seismic slowness (1/velocity) of the region between the two boreholes and the summation at a point in (τ, p) space is the travel time of a ray from $(\tau, x = 0)$ to $(px_{\max} + \tau, x_{\max})$. The transformation from (x, t) to (τ, p) space is referred to as a Radon transform. Consider now how the process might be reversed and, for the time being, concentrate on the CDP data application where t represents time.

Let $u(x, t)$ be the amplitude of the seismic signal at offset x and time t. The spatial Fourier transform:

$$u(x) = \frac{1}{2\pi}\int_{k\,\min}^{k\,\max} \hat{u}(k)e^{ikx}\,dk \qquad (5.12.7)$$

Since $p = t/x = 1/v$, let $k = |w|p$ then $dk = |w|\,dp$ and equation (5.12.7) becomes

$$u(x) = \frac{|w|}{2\pi}\int_{p\,\min}^{p\,\max} \hat{u}(p)e^{iwpx}\,dp \qquad (5.12.8)$$

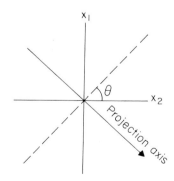

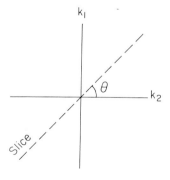

Fig. 5.16.

Then one simply carries out a two-dimensional inverse Fourier transform to obtain the original velocity field. However, problems can arise when the data in wave number space are interpolated from polar coordinates to the constant sample interval in cartesian coordinates required for the final 2-D inverse Fourier transform. The number of data points per unit area decreases with distance from the origin. So interpolation error will inevitably increase for larger values of wave number

The two-dimensional Fourier transform is then

$$u(x, t) = \frac{1}{4\pi^2}\int_{w\,\min}^{w\,\max}\int_{p\,\min}^{p\,\max} \hat{u}(w, p)|w|e^{-iw(t - px)}$$
$$\cdot\,dp\,dw \qquad (5.12.9)$$

$$= \frac{1}{2\pi}\int_{p\,\min}^{p\,\max} \hat{\hat{u}}(t - px, p) * (\text{F.T.}^{-1}|w|)\,dp$$

$$(5.12.10)$$

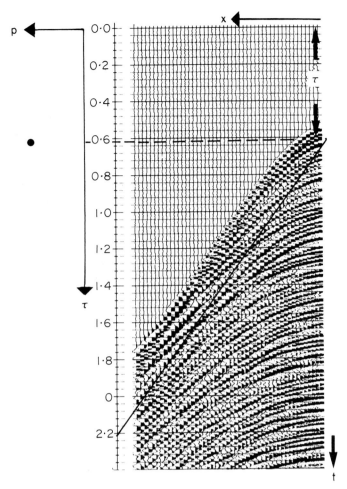

Fig. 5.17. The convolutional method (or filtered back-projection) of tomographic inversion is the same process as inverse τ-p inversion. The figure shows how a slant stack of CDP seismic data transforms to a point in τ-p space.

$\hat{u}(t - px, p)$ is the Radon transform of $u(x, t)$. Now

$$\int_{p\,min}^{p\,max} \hat{u}(t - px, p)\, dp$$

represents summation in (τ, p) space along a straight line with intercept $\tau = t$ and gradient $d\tau/dp = -x$. So the reverse process from (τ, p) space to (t, x) space is very nearly the same operation as the forward transform. The only difference is that each column in (τ, p) space must first be convolved with the inverse transform of $|w|$; hence the name 'filtered' back-projection. For further details see Robinson (1982) and Scudder (1978).

Fig. 5.18 shows a synthetic example of this process. In Fig. 5.18a one can imagine that the left-hand edge of the field represents a vertical borehole containing shots and the right hand edge another vertical borehole containing receivers. The arrows show the maximum and minimum slopes of the rays that pass across the field. Fig. 5.18b is the Radon or (τ-p) transform of Fig. 5.18a, and Fig. 5.18c is the inverse transform of Fig. 5.18b calculated using equation (5.12.10). As might be expected, the vertical edges of the L shape are poorly resolved since there are no rays that pass vertically or even near vertically from the top to the bottom of the

image. In the absence of complete azimuthal coverage, the image is blurred in the direction of the maximum density of rays.

The relationship between the Fourier transform and convolutional methods described in these last two sec-

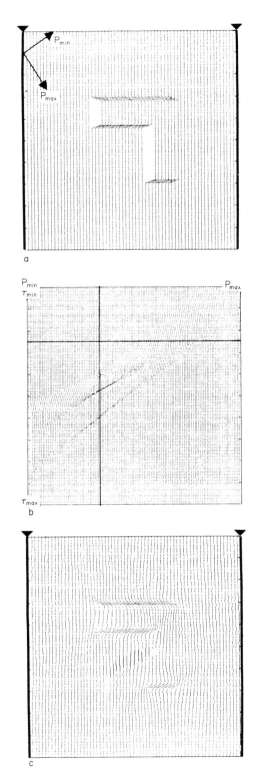

Fig. 5.18. (a) A hypothetical 'L' shaped velocity anomaly between two vertical boreholes. p_{min} and p_{max} are the maximum and minimum angles that rays leave the left-hand borehole; (b) The τ-p transform of (a); (c) The inverse transform of (b). Note that the vertical edges of the 'L' shape are blurred since the ray coverage is from the left and right boundaries of the region only.

tions is shown diagrammatically in Fig. 5.19. One proceeds from the image space to the projection space via the Radon transform. In slant stack terminology, the image space is the gathered seismic data and the projection space is the (τ, p) transform or slant stack of these data. In imaging terminology, the image space is the velocity or attenuation field that one wishes to determine and the projection space is the observed data of the travel times or amplitudes for a range of ray propagation directions.

The Fourier space is simply the 2-D Fourier transform of the image space, whereas the projection space is linked to the Fourier space via a series of 1-D Fourier transformations. Finally, one proceeds from the projection space directly to the image space via the inverse Radon transform which is convolutional back projection.

In other words, rays are traced through the region to be imaged and travel times (or amplitudes) calculated. The difference between the observed and calculated times (or amplitudes) is re-distributed back along each ray path and new values for x are calculated according to equations (5.12.11). Rays are traced again through the revised velocity field and the process is repeated. Note that if cell j is not intersected by any rays, then the value of x_j is left unchanged.

A very large number of variations on this general theme have been proposed and the reader is referred to the following references for details: Dines and Lytle, 1979; Gordon, 1974; Herman et al., 1973; Scudder, 1978; Lytle and Dines, 1980. The algorithm used in the following example is described in Mason (1981).

Fig. 5.20 shows the results of a simple synthetic tomographic experiment, which is supposed to rep-

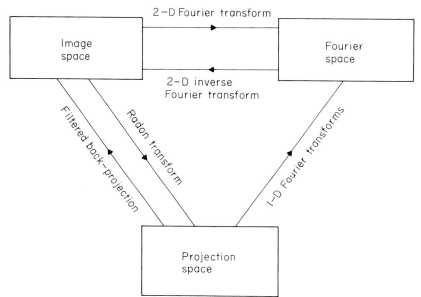

Fig. 5.19. Relationship between image space, projection space and Fourier space.

5.12.4 Algebraic reconstruction techniques (ART)

To summarise what has gone before, the advantages of matrix inversion are that any source–receiver array can be handled easily and one is not restricted to a simple form of ray path. The disadvantage is that the method is slow, sometimes quite unmanageably slow. Fourier transform and filtered back-projection methods are fast, but inflexible with regard to geometry and assumed ray path. With ART to a certain extent, one obtains the best of both worlds.

The starting point for ART is the same matrix equation as in Section 5.12.1. The only difference is that an iterative solution is sought. The original ART algorithm is defined as follows: given m data in vector y

$$y = Ax$$

where A is an $m \times n^2$ matrix where $(n \times n)$ are the dimensions of the region to be imaged.

Define calculated data $y_i^q = \sum_j A_{ij} x_j^q$ for each ray i and each iteration q. Then

$$x_j^{q+1} = x_j^q + A_{ij} \frac{(y_i - y_i^q)}{\sum_j A_{ij}^2} \qquad (5.12.11)$$

resent the use of cross-borehole tomography for imaging local structures, in this example a faulted bed. The velocity contrast between the bed and the surrounding medium is 10 per cent.

Although the rays are appreciably bent, the inversion was accomplished assuming that all the rays were completely straight. This assumption dramatically reduces the computational effort compared to a full curvi-ray inversion, and as can be seen the result is quite acceptable. Obviously, for large velocity contrasts, the straight ray approximation will give unacceptable results. However, as a rule of thumb, velocity contrasts of up to 15 per cent are tolerably well handled by straight ray inversion.

In the preceding discussion, it has been tacitly assumed that the wave propagation is adequately described in terms of rays. If the wavelength of the energy is of the same order as the structures to be imaged, then this is a highly dubious assumption. In such cases it is necessary to devise some tomographic procedure that allows for diffraction effects and, if possible, takes into account the full waveform of the transmitted energy and not just the travel time or amplitude

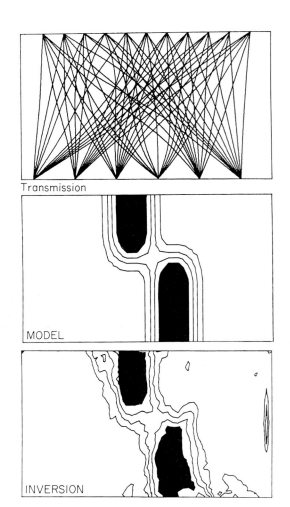

Fig. 5.20. A synthetic example of ART inversion showing the ray coverage, velocity model (10 per cent velocity anomaly) and resulting tomographic inversion. The inversion was accomplished assuming that all rays were straight.

of the earliest arriving phase. This is a very active area of current research and some recent developments will be only briefly alluded to here.

Starting with the assumption that the inhomogeneities to be imaged are only weakly scattering it can be shown that an approximate solution of the scalar wave equation leads to a modification of the projection slice theorem described in Section 5.12.2. The Fourier

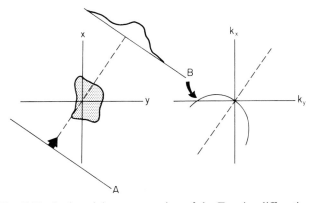

Fig. 5.21. A pictorial representation of the Fourier diffraction projection theorem. The Fourier transform of the forward scattered wave field, B, from an incident plane wave, A, gives the values of the two-dimensional Fourier transform of the object along a circular arc as shown.

diffraction projection theorem states that the Fourier transform of the forward scattered wave field measured on a line perpendicular to the direction of propagation of the wave gives the values of the 2-D Fourier transform of the object along a circular arc (as in Fig. 5.21) whose radius, k_0, is equal to $(2\pi/\lambda)$ where λ is the wavelength of the wave field in the background medium. Hence the 2-D Fourier transform of the region to be imaged is built up from a grid of circular arcs rather than a polar grid in the ray theoretical projection slice theorem.

The same interpolation problems described in Section 5.12.2 apply and an equivalent wave theoretical approach to the method described in Section 5.12.3 has been formulated. This is referred to as 'the filtered backpropagation method'. The reader is referred to Pan and Kak (1983) and Devaney (1984) for details of these procedures. Two general references, Herman (1980) and Deans (1983) are also recommended.

5.13 Appendix: singular value decomposition of a matrix A

If A is an $(n \times m)$ matrix, then following Lanczos (1961), a symmetric matrix S can be formed

$$S = \begin{pmatrix} \mathbf{0} & \vdots & A \\ \cdots & \cdots & \cdots \\ A^T & \vdots & \mathbf{0} \end{pmatrix} \begin{matrix} \} n \\ \\ \} m \end{matrix} \qquad (5.13.1)$$

Solve for the eigenvalues, λ_i, and eigenvectors of S

$$\begin{pmatrix} \mathbf{0} & \vdots & A \\ \cdots & \cdots & \cdots \\ A^T & \vdots & \mathbf{0} \end{pmatrix} \begin{pmatrix} \mathbf{u}_i \\ \cdots \\ \mathbf{v}_i \end{pmatrix} = \lambda_i \begin{pmatrix} \mathbf{u}_i \\ \cdots \\ \mathbf{v}_i \end{pmatrix} \qquad (5.13.2)$$

The eigenvectors are partitioned into $\mathbf{u}_i$ and $\mathbf{v}_i$, which are eigenvectors associated with the columns and rows of A respectively. Hence

$$A\mathbf{v}_i = \lambda_i \mathbf{u}_i \qquad (5.13.3)$$

$$A^T\mathbf{u}_i = \lambda_i \mathbf{v}_i \qquad (5.13.4)$$

Substituting (5.13.4) into (5.13.3) and vice versa one obtains

$$AA^T\mathbf{u}_i = \lambda_i^2 \mathbf{u}_i \qquad (5.13.5)$$

$$A^TA\mathbf{v}_i = \lambda_i^2 \mathbf{v}_i \qquad (5.13.6)$$

Since AA^T and A^TA are both Hermitian, the eigenvectors $\mathbf{u}_i$ $(i = 1, n)$ and $\mathbf{v}_i$ $(i = 1, m)$ are orthogonal and the eigenvalues are real.

If U and V are matrices whose columns are the normalised eigenvectors $\mathbf{u}_i$ and $\mathbf{v}_i$ respectively, and Λ is a diagonal matrix whose main diagonal elements are the eigenvectors λ_i, then (5.13.3) can be expressed as

$$AV = U\Lambda \qquad (5.13.7)$$

However, A is an $(n \times m)$ matrix, V is $(m \times m)$, U is $(n \times n)$ and from equation (5.13.2) Λ is $(n + m, n + m)$. So if equation (5.13.7) is to be balanced (same number of rows and columns on the left- and right-hand side), some of the $(n + m)$ eigenvalues must equal zero.

If $n > m$, then there must be at most m non-zero eigenvalues.

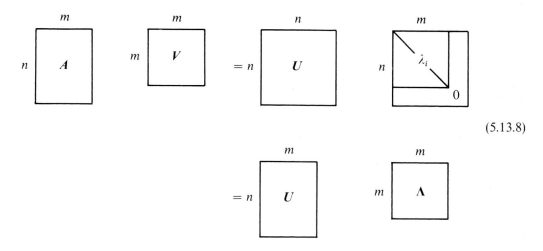

(5.13.8)

Postmultiply (5.13.8) by V^T

$$A = U\Lambda V^T \qquad (5.13.9)$$

This is referred to as the singular value decomposition of the matrix A. If $m > n$, then from equation (5.13.4) there must be at most n non-zero eigenvalues. It is quite possible that there will be fewer than m or n non-zero eigenvalues for the cases where n is greater than or less than m. For the implications of this condition, see Section 5.6.

Problems

Chapter 2

The Fourier transform

1 Using equation (2.2.6), find the Fourier transform, $X(k)$, of the function

$$x(t) = \begin{cases} A & 0 < t < 2a \\ 0 & \text{elsewhere} \end{cases}$$

$X(k)$ is complex

$$X(k) = |X(k)|e^{i\phi(k)}$$

where $|X(k)|$ is the amplitude spectrum and $\phi(k)$ is the phase spectrum.

Obtain expressions for the amplitude and phase spectra of $x(t)$. What would be amplitude and phase spectra for the functions:

(a) $\quad x(t) = \begin{cases} A & -a < t < a \\ 0 & \text{elsewhere} \end{cases}$

(b) $\quad x(t) = \begin{cases} A & a < t < 3a \\ 0 & \text{elsewhere.} \end{cases}$

Note that shifting the time origin will not change the amplitude spectrum. But note how the phase spectrum is affected.

2 If $x(t)$ is real, and $X(k)$ is its Fourier transform, show that

$$X(-k) = X^*(k)$$

where * denotes the complex conjugate of $X(k)$. This is an important result when considering the Fourier transform of the cross-correlation of two functions (see below).

3 If $x(t)$ is real, show that $|F(k)|$ is an even function of k and $\phi(k)$ is an odd function of k.

4 If a signal plus an echo is represented as

$$S(t) = f(t) - \alpha f(t - \tau)$$

show that the amplitude spectrum of $S(t)$ is the amplitude spectrum of $f(t)$ times $|H(k)|$

$$|H(k)| = [(1 - \alpha \cos k\tau)^2 + (\alpha \sin k\tau)^2]^{1/2}$$

5 Find the first frequency other than DC at which the 'ghost' spectrum is zero, when a shot is fired at a depth of 7.5 metres in water in which the speed of sound is 1500 m/sec.

Convolution

6 Following Section 2.4.2, prove the convolution theorem in its inverse form, viz., convolution in the frequency domain is equivalent to multiplication of the respective time domain functions.

7 If the digital representation of a wavelet, w, has sample values (2, 1, 0) and the earth's acoustic impedance function, e, is (1, −.5, 0, 0, 1, 0, 0, −1, 0), compute $g = w * e$, the convolution of w with e.

8 Two digital signals with sample lengths of 80 and 200 samples respectively are to be convolved together using a frequency domain algorithm. Noting that the algorithm requires sample lengths to the power of 2, how many zeros should be added to the signals to avoid possible problems due to circularity?

9 If a signal $f_i(t)$ is passed through its 'matched filter' $m(t)$, the phase spectrum of $f_0(t)$ is reduced to zero and its amplitude spectrum $|F_0(w)|$ becomes $|F_i(w)|^2$.

$$f_0(t) = f_i(t) * m(t)$$

If $(a + ib)$ is a frequency component of $F_i(w)$, and $(x + iy)$ a frequency component of $m(w)$, determine expressions for x and y in terms of a and b.

You should have proved that $M(w) = F_i^*(w)$ where * denotes the complex conjugate. Hence $F_0(w) = F_i(w)F_i^*(w) = |F_i(w)|^2$ and $\phi(w) = 0$.

Filtering

10 Referring to Fig. 2.15, note that q is an index and not strictly a frequency. For a discrete Fourier transform of a time series consisting of 64 samples with a sample interval of 2 msecs, what frequency in Hz does index $q = 27$ correspond to?

11 The db per octave slope is linear on a log-log plot of the filter's amplitude response (base 2 on the abscissa and base 10 on the ordinate). Given that the equation of the slope is

$$g' = \alpha f' + \beta \quad (\alpha, \beta \text{ constants})$$

in the log-log domain (f', g') (i.e. $f' = \log_2 f$, and $g' = \log_{10} g$, where g is the amplitude), derive the equation for the filter slopes in the linear-linear domain (f, g) (where the slope is of course no longer linear).

12 For a low-pass filter with a slope of 72 db per octave and a cut-off frequency of 64 Hz, at what frequency is the filter's amplitude spectrum 36 db down on the cut-off frequency?

13 For the same filter as given in question 12, what is the amplitude value of the filter in db at 90 Hz, if it is 1 in the pass-band of the filter?

The Z-transform

14   Find the equivalent minimum phase wavelet to the wavelet
(a)   (4, 8, 3)
(b)   (0, 1, 2).
15   Find the equivalent maximum phase wavelet of the wavelet (1, 0, .25). Factorise using complex numbers, and note that $|i| = 1$.
16   Express the inverse of the dipole (1, .2) as an infinite series. Investigate how accurately the first 4 terms of this series inverts the dipole.
17   Express the inverse of the wavelet (0, 1, .2) as an infinite series. Compare the result with the infinite series in question 16. Be sure that you appreciate why this filter is referred to as non-causal.

Auto- and cross-correlation

18   Using the theory from Section 2.7, determine the auto-correlation function of

$$x_1 = A \sin \left(\frac{2\pi l}{N} + \phi \right)$$

where ϕ is a phase factor.
19   Bearing in mind the relationship between correlation and convolution, obtain a frequency domain relationship for the Fourier transform of the cross-correlation of two functions (see question 2.) Hence show that the Fourier transform of the auto-correlation of a function is a real function.
20   Find the cross-correlation function of

$$f_1(t) = A \sin(w_1 t + \phi_1)$$

with

$$f_2(t) = B \sin(w_2 t + \phi_2)$$

where $w_1 \neq w_2$. (Think in the frequency domain and you have ten seconds!)
21   If

$$a = (3, 4, 2, -1, 1)$$

$$b = (1, 2, 1)$$

obtain the cross-correlation of a with b for positive and negative lag and the auto-correlation of a and b for positive and negative lag.
22   Prove that, for a broad band input, the cross-correlation of the input and output of a linear filter is approximately equal to the impulse response of the filter. What is the result if the input is frequency band limited?

Wiener filtering

23   If the input to a linear filter is $x = (2, 1)$, and the desired output is $s = (0, 1, 2, 1)$, determine the impulse response of the filter $h = (h, h_2, h_3)$. Construct a matrix equation of the form of equation (2.8.9). Note that $X^T X$

is the auto-correlation matrix of the input. Obtain the inverse of $X^T X$ and solve for H.
24   Convolve x with h from the previous question and compare with the desired result of (0, 1, 2, 1).
25   Following the method in Section 2.8.3.3, construct a prediction error filter to remove the echo from the signal s

$$s = (2, 1, 0, 0, 0, 1, 0.5, 0)$$

that occurs at a prediction distance of 5 samples. Convolve the result with s.

Two-dimensional Fourier transform

26   An event on a seismic time section has a dip of 15.625 msec per trace separation. Calculate the temporal frequency above which the event is spatially aliassed.
27   A surface wave with a maximum frequency of 100 Hz has an apparent velocity across an array of geophones of 200 m/s. At what distance interval should the geophones be placed if spatial aliassing of this signal is to be avoided?

Chapter 3

Geometry workshop

1   Construct a stacking chart for the following acquisition configurations, assuming a line 200 m long:
(a) 6 channel, 25 m receiver group spacing, 25 m shot pull-up.
(b) 6 channel, 25 m receiver group spacing, 6.25 m shot pull-up.
   What are the CMP spacing and fold of coverage in each case?
2   Construct a stacking chart after each of vertical and horizontal summation for the following acquisition configurations:
(a) 6 channel, 25 m receiver group spacing, 12.5 m shot pull-up.
(b) 6 channel, 25 m receiver group spacing, 6.25 m shot pull-up.
   What are the CMP spacing and fold of coverage in each of the four cases?
3   What would be the effect on fold of coverage if channel 5 was inoperative for the following:
(a) 12.5 m shot pull-up.
(b) 25 m shot pull-up.
(c) 25 m shot pull-up followed by an adjacent trace sum in processing.
4   Occasionally it is helpful to use a surface stacking chart in which the positions of receiver groups for each shot are plotted rather than reflection mid-points. Construct such a chart for the following configuration:
6 channel, 25 m receiver group spacing, 12.5 m shot pull-up.
   Indicate common shot, common depth and common offset gathers.

Chapter 5

Inverse theory

1 As the first stage in a gravity survey, you are attempting to set up four new gravity stations at convenient, but widely spaced, locations around the survey area. You wish to determine values of g at the three stations (2, 3, 4) assuming that you know the value of g_1.

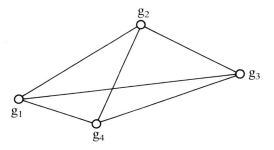

Due to the drift, measuring error, etc., your data consists of a number of values of $\Delta g_i = (g_i - g_j)$ all of which will be in error to some degree.

$$\Delta g_{ij} = g_i - g_j + e_i$$

where e_i is the error associated with the ijth measurement. Construct the matrix A of the equation

$$y = Ax$$

where

y is the data vector

x is the values of g to be determined.

What must you do to ensure that the matrix $A^T A$ is non-singular? Calculate $A^T A$ assuming that the value of g_1 is also unknown and show that $\det[A^T A] = 0$.

2 Referring to Section 5.4, determine an expression for the ith row of the matrix A' in equation (5.4.9), if in equation (5.4.10)

$$K_i(z) = \begin{cases} C & a \le z \le b \\ 0 & \begin{cases} 0 \le z < a \\ b < z \le Z \end{cases} \end{cases}$$

and $p(z)$ is expressed as the first n terms of a Fourier cosine series expansion.

$$p(z) = \sum_{i=1}^{n} (a_n \cos nz)$$

3 Solve the following set of linear equations using generalised matrix inversion

$$x_1 + 2x_3 = 1$$
$$2x_1 + 4x_3 = 2$$
$$x_2 = 1$$

In matrix notation $y = Ax$,

$$y^T = (1, 2, 1), \quad x^T = (x_1, x_2, x_3)$$

$$A = \begin{pmatrix} 1 & 0 & 2 \\ 2 & 0 & 4 \\ 0 & 1 & 0 \end{pmatrix}$$

A can be factored into the product

$$A = U\Lambda V^T \quad \text{(see Section 5.13)}$$

Obtain U and V from the equations

$$AA^T u = \lambda^2 u$$
$$A^T A v = \lambda^2 v$$

Remember to normalise the eigenvectors.

The generalised inverse operator $H = V\Lambda^- U^T$ once the zero eigenvalues have been deleted from Λ and the corresponding eigenvectors have been deleted from U and V. In this example one of the eigenvalues is zero. Reduce Λ, U and V accordingly and determine H. Hence obtain an estimate of x, since $x = Hy$.

Determine the resolution matrix VV^T where V is the (3×2) matrix of normalised eigenvectors associated with the two non-zero eigenvalues. Explain the result.

A similar example to this is presented in full in Aki and Richards (1980, pp. 679–689).

4 Consider the application of the generalised inverse approach to the very simple problem of the gravity anomaly over a buried cylinder. The gravity anomaly, A

$$A = \frac{2\pi GR^2 \sigma z}{x^2 + z^2}$$

$$\sigma = \text{density contrast}$$

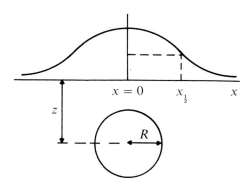

A fundamental non-uniqueness exists between the radius of the cylinder R and the density contrast σ, for any given gravity anomaly. Whereas the depth z, can be estimated from the half-width formula $z = x_{1/2}$ quite independently of any knowledge about the values of R and σ.

$$A_{\max} = \frac{2\pi GR^2 \sigma}{z} \quad (x = 0)$$

$$A_{1/2} = \frac{2\pi GR^2 \sigma z}{x_{1/2}^2 + z^2}$$

$$A_{\max} = 2A_{1/2}$$

$$\frac{1}{z} = \frac{2z}{x_{1/2}^2 + z^2}; \quad z = x_{1/2}$$

(a) Linearise the problem about some starting model (R_1, σ_1, Z_1). Obtain expressions for the columns of the matrix that transforms perturbations to the data into perturbations to the parameters, i.e. determine

$$\frac{\partial A}{\partial R}, \frac{\partial A}{\partial \sigma}, \frac{\partial A}{\partial z}$$

(b) Straightaway you should observe that the matrix will be singular—two of the columns are proportional to one another. What is the proportionality constant?

(c) The data below shows the singular value decomposition of the partial derivatives matrix formed by linearising about a starting model $R_1 = 1.0$, $\sigma = 1.0$, $Z = 10.0$ (arbitrary units). Note that the smallest eigenvalue only differs from zero due to numerical error. Hence all the eigenvalues are treated as finite and the resolution matrix is an identity matrix.

What result would you expect from an inversion with three degrees of freedom?

Form the resolution matrix, having eliminated the smallest eigenvalue and its associated eigenvector. Explain the result that you get.

Original matrix

	Col 1	Col 2	Col 3
1	0.1980	0.0990	−0.1860
2	0.1600	0.0800	−0.1200
3	0.1000	0.0500	−0.045

	Singular values	Normalised
1	3784.05772307	1.0000E 00
2	365.38629980	9.6559E 02
3	0.00000000	3.7453E-13

U matrix—orthonormal row eigenvectors

	Col 1	Col 2	Col 3
1	−0.76	−0.52	0.39
2	−0.57	0.25	−0.78
3	−0.31	0.82	0.49

V matrix—orthonormal column eigenvectors

	Col 1	Col 2	Col 3
1	−0.72	0.53	0.45
2	−0.36	0.26	−0.89
3	0.59	0.81	−0.00

5 Derive equations (5.6.16) and (5.6.17) in Section 5.6.

6 Solve the set of equations

$$\begin{pmatrix} 1 & 1 & 0 \\ 0 & 0 & 1 \\ 0 & 0 & -1 \end{pmatrix} \begin{pmatrix} x_1 \\ x_2 \\ x_3 \end{pmatrix} = \begin{pmatrix} 1 \\ 2 \\ 1 \end{pmatrix}$$

Using damped least squares, i.e. calculate

$$[A^T A + e^2 I]^{-1} A^T \begin{bmatrix} 1 \\ 2 \\ 1 \end{bmatrix} = \begin{pmatrix} x_1 \\ x_2 \\ x_3 \end{pmatrix}$$

for an arbitrary small value of e and neglect terms of order greater than e^2.

7 Referring to Section 5.9, how would equation (5.9.3) need to be modified if the reflecting interface that gives rise to the upgoing wave dips at an angle of 10 degrees?

8 In the figure below assume that rays travel in straight lines from shot positions S1 and S2 to receiver positions R1 and R2.

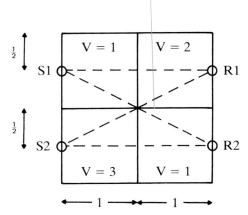

The medium between the shots and receivers consists of four square regions with velocities of 1, 2, 3, 1.

(a) Construct a matrix equation

$$t = As$$

where t contains the travel times of the four rays and s contains the slowness of the four regions.

(b) Use the algorithm defined in equation (5.12.11) to perform one or two iterations of an ART inversion, taking as your starting model a homogeneous velocity of 1.

References

Adby, P. R. and Dempster, M. A. H. (1974) *Introduction to Optimisation Methods*. Chapman and Hall, London.

Aki, K. and Richards, P. G. (1980) *Quantitative Seismology, Theory and Methods*. Vols. I and II. W. H. Freeman, San Francisco.

Balch, A. H. and Lee, M. W. (1984) *Vertical Seismic Profiling*. IHRDC, Boston.

Berkhout, A. J. (1980) Seismic migration: imaging of acoustic energy by wave field extrapolation. In: *Developments in Solid Earth Geophysics* 12. Elsevier, Amsterdam.

Berkhout, A. J. (1984a) Seismic resolution. In: *Handbook of Geophysical Exploration* 12. Geophysical Press, London/ Amsterdam.

Berkhout, A. J. (1984b) Seismic migration. In: *Developments in Solid Earth Geophysics* 14B. Elsevier, Amsterdam.

Bracewell, R. N. (1978) *The Fourier Transform and its Applications* (2nd edn). McGraw-Hill, New York.

Burg, J. P. (1967) *Maximum Entropy Spectral Analysis*. Paper presented at 37th SEG Meeting, Oklahoma, 31 October.

Capon, J. (1969) High-resolution frequency-wave-number spectrum analysis. *Proc. I.E.E.E.* **57**, 1408–1418.

Cassell, B. (1984) Vertical seismic profiles—an introduction. *First Break* **2**, No. 11, 9–19.

Chatfield, C. 1975. *The Analysis of Time Series: Theory and Practice*. Chapman and Hall, London.

Claerbout, J. F. (1976) *Fundamentals of Geophysical Data Processing*. McGraw-Hill, New York.

Claerbout, J. F. (1985) *Imaging the Earth's Interior*. Blackwell Scientific Publications, Oxford.

Claerbout, J. F. and Muir, F. (1973) Robust modelling with erratic data. *Geophysics* **38**, 826–844.

Deans, S. R. (1983) *The Radon Transform and Some of its Applications*. Wiley-Intersciences Publications, New York.

Devaney, A. J. (1984) Geophysics diffraction tomography. *I.E.E.E. Trans. Geosci. Remote Sensing* **GE-22**, 3–13.

Dines, K. A. and Lytle, R. J. (1979) Computerised geophysical tomography. *Proc. I.E.E.E.* **67**, 1065–1073.

Evenden, B. S., Stone, D. R. and Anstey, N. A. (1971) *Seismic Prospecting Instruments*. Vol. I, Geoexploration Monographs. Geopublication Associates Gebiüder Borntraeger, Berlin/Stuttgart.

Fromm, G. and Helbig, K. (1984) *Static Corrections*. Continuing Education for Explorationists, EAEG Series. International Science Services, Couwenhoven, The Netherlands.

Gazdag, J. (1978) Wave equation migration with the phase shift method. *Geophysics* **43**, 1342–1351.

Gibson, B. and Larner, K. (1984) Predictive deconvolution and the zero-phase source. *Geophysics* **49**, 379–397.

Gilbert, F. (1972) Inverse problems for the earth's normal modes. In: *The Nature of the Solid Earth*, ed. E. C. Robertson. McGraw-Hill, New York, 125–146.

Gordon, R. (1974) A tutorial on ART. *I.E.E.E. Trans. Nucl. Sci.* **NS-21**, 78–93.

Grieg, D. M. (1980) *Optimisation*. Longman, London.

Hatton, L. (1980) *Image-Rays and the Treatment of Inhomogeneity in Migration*. Paper presented at 5th SE Asia Petroleum Society Meeting, 31 March.

Hatton, L. (1981) *Aliasing and the Transient Signal*. 43rd Meeting of the EAEG, Venice, Italy.

Hatton, L. (1983a) Computer science for geophysicists I. *First Break* **1**, No. 6, 18–24.

Hatton, L. (1983b) Computer science for geophysicists II. *First Break* **1**, No. 9, 18–22.

Hatton, L. (1983c) Computer science for geophysicists III. *First Break* **1**, No. 10, 13–19.

Hatton, L. (1983d) Computer science for geophysicists IV. *First Break* **1**, No. 11, 18–23.

Hatton, L. (1984a) Computer science for geophysicists V. *First Break* **2**, No. 1, 9–15.

Hatton, L. (1984b) Computer science for geophysicists VI. *First Break* **2**, No. 9, 9–17.

Hatton, L., Larner, K. L. and Gibson, B. S. (1981) Migration of seismic data from inhomogeneous media. *Geophysics* **46**, 751–767.

Herman, G. T. (1980) *Image Reconstruction from Projections. The Fundamentals of Computerised Tomography*. Academic Press, New York.

Herman, G. T., Lent, A. and Rowlands, S. (1973) ART: mathematics and applications: a report on the mathematical foundations and on applicability to real data of the algebraic reconstruction techniques. *J. Theor. Biol.* **43**, 1–32.

Hogg, R. V. and Craig, A. T. (1959) *Introduction to Mathematical Statistics*. Macmillan, New York.

Hood, P. (1981) *Migration in Developments in Geophysical Exploration Methods* 2, ed. A. A. Fitch. Applied Science Publishers, London, 151–230.

Hosken, J. W. J. (1981) Imaging the earth's subsurface with seismic reflections. In: *The Solution of the Inverse Problem in Geophysical Interpretations*, ed. R. Cassinis. Plenum Press, New York, 179–210.

Hubral, P. (1977) Time migration—some ray theoretical aspects. *Geophysical Prospecting* **25**, 738–745.

Hubral, P. and Krey, T. (1980) *Interval Velocities from Seismic Reflection Time Measurements*. SEG Publications, Tulsa, Oklahoma.

Jackson, D. D. (1972) Interpretation of inaccurate, insufficient and inconsistent data. *Geophys. J. R. Astr. Soc.* **28**, 97–109.

Jackson, M. C. (1977) A classification of the snowiness of 100 winters—a tribute to the late L. C. W. Bonacina. *Weather* **32**, 91–97.

Johnson, C. E. (1972) *Regionalised Earth Models from Linear Programming Methods*. M.Sc. Thesis, Massachusetts Institute of Technology.

Kanasewich, E. R. (1981) *Time Sequence Analysis in Geophysics* (3rd edn). University of Alberta Press.

Kendall, M. G. and Stuart, A. (1966) *The Advanced Theory of Statistics* 3. Charles Griffin, London.

Kjartansson, E. (1979) *Attenuation of Seismic Waves in Rocks and Applications in Energy Exploration*. Ph.D. Dissertation, Stanford University.

Kline, M. (1972) *Mathematical Thought from Ancient to Modern Times*. Oxford University Press, Oxford.

Lacoss, R. T. (1971) Data adaptive spectral analysis methods. *Geophysics* **36**, 661–675.

Lanczos, C. (1961) *Linear Differential Operators*. Van Nostrand, London.

Larner, K., Chambers, R., Yang, M., Lynn, W. and Wai, W. (1983) Coherent noise in marine seismic data. *Geophysics*

48, 854–886.

Larner, K., Gibson, B. and Rothman, D. (1980) *Trace Interpolation and the Design of Seismic Surveys*. Paper presented at 50th SEG Meeting, Houston, November.

Larner, K. and Hatton, L. (1976) *Wave Equation Migration: Two Approaches*. Offshore Technology Conference OTC-2568, Houston.

Larner, K. L., Hatton, L. and Gibson, B. S. (1981) Depth migration of imaged time sections. *Geophysics* **46**, 734–750.

Lerwill, W. E. (1981) The amplitude and phase response of a seismic vibrator. *Geophysical Prospecting* **29**, 503–528.

Lighthill, M. J. (1962) *Introduction to Fourier Analysis and Generalised Functions*. Cambridge University Press, Cambridge.

Lines, L. R. and Clayton, R. W. (1977) A new approach to vibroseis deconvolution. *Geophysical Prospecting* **25**, 417.

Loewenthal, D., Lu, L., Roberson, R. and Sherwood, J. W. C. (1976) The wave equation applied to migration. *Geophysical Prospecting* **24**, 380–399.

Loveridge, M. M., Parkes, G. E., Hatton, L. and Worthington, M. H. (1984) Effects of marine source array directivity on seismic data and source signature deconvolution. *First Break* **2**, No. 7, 11–23.

Lytle, R. J. and Dines, K. A. (1980) Interative ray-tracing between boreholes for underground image reconstruction. *I.E.E.E. Trans. Geosci. Remote Sensing* **GE-18**, 234–239.

Mason, I. M. (1981) Algebraic reconstruction of a two-dimensional velocity inhomogeneity in the High Hazles seam of Thoresby colliery. *Geophysics* **46**, 298–308.

Mersereau, R. M. and Oppenheim, A. V. (1974) Digital reconstruction of multi-dimensional signals from their projections. *Proc. I.E.E.E.* **62**, 1319–1338.

Neidell, N. S. and Taner, M. T. (1971) Semblance and other coherency measures for multichannel data. *Geophysics* **36**, 482–497.

Newland, D. E. (1975) *An Introduction to Random Vibrations and Spectral Analysis*. Longman, London.

Newman, P. (1973) Divergence effects in a layered earth. *Geophysics* **38**, 481–488.

Newman, P. (1975) *Amplitude and Phase Properties of a Digital Migration Process*. Paper presented at 37th Meeting of the EAEG, Bergen, Norway.

Newman, P. (1984) Seismic response to sea-floor diffractors. *First Break* **2**, No. 2, 9–19.

Noble, B. and Daniel, J. W. (1977) *Applied Linear Algebra* (2nd edn). Prentice-Hall, Englewood Cliffs, N.J.

Oppenheim, A. V. and Schafer, R. W. (1975) *Digital Signal Processing*. Prentice-Hall, Englewood Cliffs, N.J.

Oristaglio, M. (1978) *Geophysical Investigations of Earth Structure within the Vicinity of Boreholes*. D.Phil. Thesis, Oxford University.

Pan, S. X. and Kak, A. C. (1983) A computational study of reconstruction algorithms for diffraction tomography: interpolation versus filtered back propagation. *I.E.E.E. Trans. Acoustics, Speech and Sig. Proc.* **ASSP-31**, 1262–

1275.

Parkes, G. E. and Hatton, L. (1984) *Toward a Systematic Understanding of the Effects of Velocity Model Errors on Depth and Time Migration of Seismic Data*. Paper presented at 54th SEG Meeting, Atlanta, December.

Parkes, G. E., Ziolkowski, A., Hatton, L. and Haugland, T. (1984) The signature of an airgun array: computation from near field measurements including interactions—practical considerations. *Geophysics* **48**, 105–111.

Robinson, E. A. (1982) Spectral approach to geophysical inversion by Lorentz, Fourier, and Radon transforms. *Proc. I.E.E.E.* **70**, 1039–1054.

Robinson, E. A. (1983) *Migration of Geophysical Data*. Reidel, Dordrecht.

Robinson, E. A. and Treitel, S. (1980) *Geophysical Signal Analysis*. Prentice-Hall, Englewood Cliffs, N.J.

Schultz, P. S. and Sherwood, J. W. C. (1980) Depth migration before stack. *Geophysics* **45**, 376–393.

Scudder, H. J. (1978) Introduction to computer-aided tomography. *Proc. I.E.E.E.* **66**, 628–637.

Seeman, B. and Horowicz, L. (1983) Vertical seismic profiling: separation of upgoing and downgoing acoustic waves in a stratified medium. *Geophysics* **48**, 555–568.

Society of Exploration Geophysicists (1908) *Digital Tape Standards*. SEG Publications, Tulsa, Oklahoma.

Stainsby, S. D. (1984) *Seismic Attenuation and VSP Processing*. D.Phil Thesis, Oxford University.

Stolt, R. H. (1978) Migration by Fourier transform. *Geophysics* **43**, 23–48.

Taner, M. T., Koehler, F. and Sheriff, R. E. (1979) Complex seismic trace analysis. *Geophysics* **44**, 1041–1063.

Ursin, B. (1978) Attenuation of coherent noise in marine seismic exploration using very long arrays. *Geophysical Prospecting* **26**, 722–747.

Vajda, S. (1970) *The Theory of Games and Linear Programming*. Monograph Science Paperback SP50. Methuen, London.

Waters, K. H. (1978) *Reflection Seismology*. Wiley-Interscience, New York.

White, J. E. (1983) *Underground Sounds*. Methods in Geochemistry and Geophysics 18. Elsevier, Amsterdam.

Wiggins, R. A., Larner, K. L. and Wisecup, R. D. (1976) Residual statics analysis as a general linear inverse problem. *Geophysics* **41**, 922–938.

Worthington, M. H., Cleary, J. R. and Anderssen, R. S. (1974) Upper and lower mantle shear velocity modelling by Monte Carlo inversion. *Geophys. J. R. Astr. Soc.* **36**, 91–104.

Ziolkowski, A. (1984a) The Delft airgun experiment. *First Break* **2**, No. 6, 9–18.

Ziolkowski, A. (1984b) *Deconvolution*. IHRDC, Boston.

Ziolkowski, A., Parkes, G., Hatton, L. and Haughland, T. (1982) The signature of an airgun array: computation from near field measurements including interactions. *Geophysics* **47**, 1413–1421.

Index

Page numbers in italics refer to figures, numbers in bold to tables.

Absorption 57
Acoustic impedance 17, 30, 51, 54
Acoustic pingers, use of 121
Acoustic velocity profile 52
AGC *see* automatic gain control
Air-gun arrays 80
Algebraic reconstruction techniques (ART),
    tomography 161–2
Aliassing 13–14, *14*, 35, 97, 121
    benign 137
    severe 137
    spatial 42–6, 93–4
        and migration 136–7
Aliassing and de-aliassing in 1-D and 2-D 41–2
All-anticipation component filter 25
All-memory component filter 25
American Standard Code for Information Interchange *see*
    ASCII code
Amplitude 72
    and data discontinuities 94
Amplitude abnormalities *see* random noise attenuation
Amplitude decay 56–7, 64
Amplitude equalisation 105
Amplitude errors, diagnosis of 129
Amplitude level correction 105
Amplitude monitoring 110
Amplitude spectrum 77
    selection of 74–5
Analogue filtering 13
Analogue plotter 6
Analytic signal 25–6
Anti-alias filtering 21
Appraisal surveys 106
Array processors 6–8, 47
Array simulation 104–5
ASCII code 8, 50
Attenuation techniques, general 92
Auto-correlation 106
    and periodicity 91
    and spectral analysis 34
    zero-lag 82
Auto-correlation function 26–31
Auto-correlation lag, minumum and maximum *82–3*, 85–6
Auto-correlation window position and length 86
Automated migration *see* hyperbolic summation
Automatic gain control 59, *63*
Automatic gain control normalisation 72
Average velocity 64
Averaging 52

Background noise, constant 87, 89
Band-pass filtering 86, 105
    single channel 87
Band-pass filters 19–20, 21, 111
Bartlett spectral windows 36, *37*

Bias 101–2
Bin centre 118–19
Bin, definition 118
Bin dislocation 119
Binning and the crooked line 117–19
Bit-reversing 15
Boreholes and sonic data 152
Broad band filters 110, 111
Brute stack 107, 110
Buffer memory, larger size needed 5
Butterflies 15
Byte organisation 3

Cable positioning, a major problem 121
Cache memory 3
Causality and inversion 25
Central processing unit 1–2, 7
Character format 8
Check shot survey 152
Circular convolution 19
Circularity 18
Clip level 102
CMP assumption, breakdown of 109
CMP data 52–3
CMP stack and random noise attenuation 87
Coherency filters, *f–k 94, 95, 96*
Coherent energy 55
Coherent noise 55, 89, *95*
Coherent noise attenuation 86
Coherent noise rejection filtering 105, 108
Coloured light 151
Combination displays 68
Common mid-point and bins 118
Common mid-point data (multi-coverage) 52–3
Common mid-point stacking 61–4
Common sense, use of 129
Commutativity
    and migration 135–6
    with filters 135–6
    with gain 136
Complex cepstrum 82
Computer documentation, quality of 1
Computer systems and seismic data processing 1–8
Computers used in seismic data processing, relative CPU
    performance **2**
Constant velocity stacks (CVS) 66, *67*
Convergence, a problem 151
Convolution 16–19, 76
    and correlation 27
    and multiple reflections 55
    as a matrix 31
    definition 16
Convolution equation and Z-transform 22
Convolution theorem
    and circularity 18

Convolution theorem (*cont.*)
  discrete   17–18
Convolutional filtering   17
Convolutional methods, tomography   159–61
Convolutional model (extended 1-D), seismic trace   18–19
Correlation and convolution   27
Correlation functions
  definitions   26
  periodicities in   29–30
Correlations and *S/N* ratio   30–1
Covariance matrix   146
CPU *see* central processing unit
Crooked line concept (2 1/4-D processing)   116, 117–19
Cross-section display, data enhancement techniques   100–4
Cross-correlation function   26–31
Cross-correlation, normalised   68

Damped least squares   150–1, 153, 154, 155
  and residual statics   154–5
Damped least squares inversion   150
Damping and the frequency domain   151
Damping parameter   150
DAS   110–11
DAS test suite   110–11, *112–13*
Data
  and amplitude discontinuities   94
  interpretation of   47
Data compression and reconnaissance processing   52
Data dependent gain and sliding windows   104
Data enhancement techniques   72–104
  cross-section display   100–4
  random noise attenuation   87–9
  temporal resolution   72–87
  unwanted signal, attenuation of   89–100
Data error (GMI)   146–7
Data formats, internal   8
Data mixing, pre-stack   92
Data pre-conditioning   84
Data preparation before stack   70–2
Data record, SEG Y   50
Data weighting   150
Datum   121
DBS   110–11
DBS test suite   *85–6*, 110–11
Decay compensation   57–9
Decibels   19
Decimation in frequency   14
Decimation in time   14
Deconvolution   19, 25, 72
  deterministic   76–82
  multiple window   86
  predictive   47, 83, 100, 105
    and Wiener filter   33–4
    theory of   28
  single window   108
  spiking   85
  statistical, using Wiener filter   82–7
Deconvolution before and after stack   110–11
Deconvolution filters   152
Delft experiment data   79–80
Demultiplexing   1, 47–50
Depth migration   127, 128
Designature   104
Deterministic deconvolution   76–82
Deterministic deconvolution techniques (future)   87
Deterministic methods, resolution improvement   76
DFT *see* Fourier transform, discrete

Diagonal matrix   146
Differential NMO techniques, exploitation of   94, 96
Diffraction effects and topography   161
Diffraction features and salt deformation   109
Digital computers and seismic data processing   47
Digital plotters   6
Dimensionality, categories of   116
Dip filters, 2-D   93
Dip range selection   93
Dipoles   22–3, 25
Dipping layers, NMO velocity   66
Dip-reversal   42, 136
Direct arrivals (marine shot files)   60–1
Direction indicator   115
Directivity effects   71, 79
Disc access time   6
Discrete convolution theorem   17–18
Discrete Fourier transform   12–16, 18
  and time shifts   14
    example, spectrum of a ghost   15–16
  inverse   12
  relationships   13
Discrete Wiener–Hopf equation   32
Discs   5–6
Dispersion   137
Display elements and parameter selection   100–3
Display (gain) level 101, *102*
Divergence loss   56
Dix's equation   116, 139
Dominant horizon extraction   82
Dot densities and seismic plots   6
Dot mode, plotting   *100*
Dot spacing and size, digital plotters   103
Dynamite   122

Earth filtering   82
EBCDIC   8, 50
Electrostatic plotters   100–1
Elevation   121
End of file (EOF)   5
End of medium (EOM)   5
Equalisation
  data dependent   58–9
  data independent   57–8
Equalisation techniques   57–9
Equivalent layer theorem of potential theory   139
Equivalent minimum-phase wavelet   23
Errors
  disc   6
  magnetic tape   3
Evaluation surveys   106
Exploding reflector model   123, 126
Extended Binary Coded Decimal Interchange Code *see* EBCDIC

F–k (domain) filtering   93–4, 105
  and wave separation   153
  effects of   *45*
F–k (domain) multiple attenuation   97, *98*, *99*
F–k (domain) tapers   94
Far-field recording   78–9
Far-field signature   80, *81*
Fast Fourier transform   14–15, 18, 34, 94
  interpolation example   15
Fermat's principle   51
FFT *see* Fourier transform, fast

Field files 70
Field statics 121
Filter parameters 87, *88*, *90–1*
Filter slopes 21, 75
Filtered back-projection 159–61
Filtering
   2-D 39, 41, 94
   2-D convolutional 105
     definition 19
   anti-alias 21
   band-pass 86, 105
     single channel 87
   convolutional 17
   f–k domain 105
   of coherent noise 105
   time-variant 111, 114
   types of 19
Filtering 19–22
Filters
   all-anticipation component 25
   all-memory component 25
   band-pass 19–20, 21, 111
   broad band 110–111
   coherency 89
   coherency (*f–k*) *94, 95, 96*
   dip, 2-D 93
   high-pass 20
   low-pass 20
   narrow band 111, 114
   notch 20
   prediction error 33
   predictive deconvolution 33
   slopes of 21, 75
   variable, low/high cut 114
   Wiener 31–4, 76
     uses of 32–4
   Wiener inverse 23
   zero-phase 87
Final amplitude balance and display 114
Final film plotting 101
Finite-difference method 126
Finite-difference techniques 133
Floating datum 121
Floating point accelerators 6–8
Floating point format 8
Formats, multiplexed and demultiplexed 48
FORTRAN and array processors 7
Fourier cosine transform 34
Fourier diffraction projection theorem 162
Fourier theory, basic 11–12
Fourier transform
   2-D 38–46, 159
   cosine 34
   discrete 12–16, 18
     and time shifts 14
     example, spectrum of a ghost 15–16
     inverse 12
     relationships 13
   fast 14–15, 18, 34, 94
     interpolation example 15
   inverse 153
   related to Z-transform 22
   spatial 159
Fourier transform 11–12, 34, 35, 151, 153
   and convolutional methods, relationship between *161*
Fourier transform methods, tomography 158–9
Frequency and absorption 57
Frequency content, high pre-stack 92

Gain compensation *58*
Gain header constants 48
Gain recovery (seismic tapes) 47–50
Gain treatments, alternative 105–6
Gaussian humps 151
Gaussian noise 133
   and the power spectrum *37*
Gaussian noise section, unmigrated and migrated 133, *134*
Gaussian noise series 27, *28, 29*
GCR recording 3
Generalised matrix inversion 143–6
   practical problems 146–50
Geological continuity, and stacking velocity
     interpretation 70
Geometry
   and statics 121–2
   irregular and extended 116–21
Geophone amplitude data 56
Ghost reflections 54
Ghost spectra, example of DFT usage 15–16
GMI 143–6
Gradient methods 140

Head crash 5–6
Header constants, extended and general 48
Header record 48, *49*
Hedgehog inversion 140, 141, 156
Herglotz–Weichart equation 139
High-pass filters 20
Hilbert transform and the minimum-phase wavelet 23
Hilbert transformation filter 25
Horizontal (trace) sums 52
Huygen's principle 124
Hydrophones (deep-towed) and directivity effects 79
Hyperbolic summation 124

IBM character format *see* EBCDIC
IBM floating point format (SEG Y) 4, 49, 50
Image rays 127, 128–9, *130*
Information density matrix 146
Inside trace muting 96
Instrument response 81–2
Integer format 8
Interference 55
Interpolation
   example of FFT usage 15
   linear 38
   plotters 102–3
Interpolator
   quality of 15
   spectrum of 37–8
Inter-record gap 3–4, 50
Interval velocities 64, 116
Inverse Fourier transform 153
Inverse operator 144
Inverse problem, the 19
Inverse theory and applications 139–63
Inversion, definition 139
IRG *see* inter-record gap
Irregular and extended geometry 116–21
Iterative solutions 155

Kick-out 70, 71
Kick-out point 107
Kirchhoff integral and solution 124, 126, 136

Kirchhoff summation   124, 126
Kirchhoff summation formulation   132–3

Lag   26, 30, 31, 34, 54–5, 89
Land and marine processing, differences between   121–2
Laser plotters   101
Laser technology   6
Least squares inverse   151
Least squares method   31–2
Least squares minimisation criteria   32
Least squares standard solution   150
Levinson algorithm   34
Line tieing   116
Linear convolution   19
Linear interpolation   38
Linear programming   155–6
   extremal inversion   156
   L_1 norm inversion   156
Linearisation   142–3
Location map   115
Long arrays   92
   simulated   92–3
Low-pass filters   20

Magnetic compasses, magnetic field problem   121
Main memory   2
Marine seismic data processing
   common mid-point stacking   61–4
   data preparation before stacking   70–2
   fundamentals of   50–72
   mid-point assumption   50–2
   multi-coverage common mid-point (CMP) data   52–3
   nature of seismic traces   53–9
   patterns within a marine file shot   59–61
   stacking velocity generation   64–70
Marine seismic data processing schemes   104–16
   alternative gain treatments   105–6
   parameter selection philosophy   106–9
   process ordering   104–5
   quality control procedures   109–10
   relative amplitude preservation   106
   section annotation and labelling   114–16
   testing schemes   110–14
Marine seismic data processing sequence   104
Matrix formulation of the inverse problem   141–2
Matrix inversion   157–8
   disadvantage of   157–8
   generalised   143–6
      practical problems   146–50
Maximum convexity migration   124
Maximum entropy method, data-adaptive spectral analysis   36, 38
Maximum likelihood method, data-adaptive spectral analysis   36, 37
Maximum simplicity method   141
Mean amplitude stack   61–2, 64
Mean versus median (stacking)   96
Median stacking   96–7
Megaflop (unit of floating point performance)   7, 14
MEM see maximum entropy method
Method of maximum simplicity   141
Microprocessors and seismic computing systems   2
Mid-point assumption   50–2
Migrated depth domain   127–8, 128, 129
Migrated time domain   125–7, 127, 128, 129, 131

Migration before stack   138
Migration equation   124
Migration formulae   137–8
   approximate position after migration   137
   dip before and after migration   137–8
Migration (seismic)   109, 114, 121, 123–38
   2-D versus 3-D   133, 135
   and commutativity   135–6
   and noise   132–5
   and spatial aliassing   136–7
   and the 3 domains   123–7
   basic principles for a 2-D problem   123–4
   natural coordinates   127–8
   useful formulae   137–8
   velocities and velocity sensitivity   128–32
Migration velocities   116, 128–9, 132
Minimum and maximum entropy   82
Minimum-phase, concept of   23
Minimum-phase deconvolution technique   76
Minimum-phase sections   74
Minimum-phase wavelet   23
MIP (million instructions per second)   1, 2
Missed shots (normalisation)   64
Mixed phase wavelets   24
MLM see maximum likelihood method
Mode   101
Mode conversion   57
Model curvature   132
Monte Carlo inversion   140, 141, 156
Multichannel techniques, constant background noise reduction   89
Multi-element shots   78–9
Multiple attenuation   105
Multiple attenuation techniques   94–100
Multiple reflections   54–5, 71
Multiplexed formats   48
Mute selection   71–2
Muting   64, 73, 94, 108

Narrow band filters   111, 114
Natural coordinates   127–8
Near-field signature   78–81
Near trace displays   109–10
NMO   60
   and stack   86
NMO assumptions   71
NMO corrected CMP gather displays   110
NMO correction   93, 97, 98, 155
   and distortion   71
NMO differential   100
NMO equation   60, 71
   multilayer case   65–6
NMO errors   96
NMO stretch   71
NMO velocity   64, 68, 129
   dipping layers   66
Noise and its auto-correlation   29
Noise and migration   132–5
Noise bursts   89
Noise generating mechanisms   56
Noise perturbations   146
Non-linear methods (inversion)   140–1
Non-uniqueness   139–40
Normal moveout see NMO
Notch filters   20
NRZI recording   3
Nyquist circular frequency   13

Nyquist criterion 35
Nyquist frequency 13, 14, 16, 21, 38
Nyquist limitation 13
Nyquist linear frequency 13
Nyquist, spatial 33–4

Objective function 140
Octave 19
Offset mute checking 110
Offset muting 64, 71–2, 107, 111
Offset VSP 152
Operator length, Wiener filter design 77
Operator spectra 77
Operator time-zero position 78
Optical prediction theory 36

Parallelism 7
Parameter choice (statistical deconvolution) 85–6
Parameter considerations, global 86
Parameter estimates, uncertainty in 139
Parameter selection
  and display elements 100–3
  plotters 101
Parameter selection (marine seismic data processing) 106–9
  examples 107–9
Parameter weighting 147–9, 154
Parity 3
Parsimony 82
Partial energy and minimum-delay 23
Pass-band 19–20, 21
PE recording see recording techniques
Periodicity 28, 89
  and auto-correlation 111
  exploitation of 100
  in correlation functions 29–30
Phase 122
Phase and dipoles 22–3
Phase spectrum 72
  selection of 74
Pick, quality of 70
Plotters
  and parameter selection 101
  digital, dot spacing and size 103
  electrostatic (on-line) 100–1
  laser 101
  off-line 100
Plotting and plotters 6
Positioning errors, diagnosis of 129
Post-stack processes 105
Power spectrum 34, 37
Predicted trace 100
Prediction distance (lag) 34
Prediction error filter (whitening filter) 33, 36
Predictive deconvolution 47, 85, 100, 105
  and Wiener filter 33–4
  theory of 28
Pre-whitening 35
Profiles, consistency of 70
Projection slice theorem 158
  modification of 162

Quality control procedures (marine seismic data processing) 109–10

Radix-2 algorithm 15
Radon transform 159, 160, 161

Random noise 55–6
Random noise attenuation 87–9
Random noise reduction, methods of 56
Raster plotting 6, 7
Raw stack 107, 110
Raw stack display 110
Ray paths 154, 157, 161
  complex 54, 71
  contorted 109
Ray paths (mid-point) 51
Ray theory 121
Ray-theoretical migration, simple 124
Ray-trace migration 127, 133
Real time domain series 13
Receiver differences 122
Reconnaissance processing and data compression 52
Record blocking 4–5
Recording and processing (tape) 4–5
Recording techniques 3
Reflecting points, mid-point assumption 51
Reflection amplitudes and acoustic velocity 65
Reflection method 50–1
Reflection resolution 107
Reflection series 54
  recovery of 72
Reflections (marine shot files) 59–60
Refraction seismology delay-time experiment 142, 145
Refractions (marine shot files) 61
Regional surveys 106
Regularity and 3-D processing 120
Relative amplitude preservation 106, 107
Residual statics 121, 154–5
Resolution criteria 75–6
Resolution improvement, techniques for 76
Resolution matrix 146, 147
Ringing 21, 75, 77
Roll-off 52
Roll-on 52
Root mean square velocity 64

S/N ratio 30, 77, 82, 84, 86, 88, 90–1, 107, 111, 122, 151
  in migration 135
S/N ratio improvement 62
Salt diapir, parameter selection 108, 109
Salt dome model, velocity relationships 131, 132
Sampling and aliassing 13–14
Sand/shale sequence, parameter selection 108, 109
Scale 101
Scattering 57
Scatters 105
Section annotation and labelling 114–16
  acquisition details 116
  location information 115–16
  processing details 116
Sedimentary basin, parameter selection 107, 108
SEG A format 48
SEG B format 4, 48–9
  data record 48–9
  header record 48, 49
SEG C format 48
SEG D format 48
SEG (Society of Exploration Geophysicists) 47
SEG Y format 4, 48, 49, 50
SEG Y format tape, and off-line plotters 100
Seismic field tapes, problems with 4
Seismic migration as a transform 123
Seismic processing system, elements of 1–8

Seismic pulse
    ideal *54*
    modified 54, *55*
Seismic traces, nature of 53–9
Seismogram *see* seismic traces
Seismology and the 2-D transform 39
Semblance 66
Semblance values 68, *69*
Sequential recording and access time 5
Shah function 35, 42
Shear velocity 147, 156
Shear waves 57, 147
Shelf area, parameter selection 108
Shot files 59
Shot gather displays, selected 110
Shotpoint location map 115
Signal-to-noise ratio *see* S/N ratio
Signature deconvolution 104, 154
    and ghost spectra 80–1
Signature shaping and Wiener filter 33
Single element source recording, advantages and
        disadvantages 78
Single filter panel suite 110
Single fold coverage 52
Single trace amplitude analysis 57, *58*
Singular value decomposition 144, 155, 157
    of a matrix 162–3
Slant stacking 159, 161
Smoothing filters 34
Snell's law 51, 121, 124
    absence of 127
Society of Exploration Geophysicists (SEG) 47
Source component inversion 78–82
Source differences 122
Source simulation 92, 93
Spatial aliassing 42–6, 93–4
    and migration 136–7
Spatial Fourier transform 159
Spectral analysis 34–8
    and auto-correlation 27
    data-adaptive 36–7
    data-independent 36
Spectral averaging 82
Spectral broadening 82
Spectrum, convolutional model of 35
Spherical divergence correction 106
Spherical divergence formulae 105
Spikes 89
Spiking deconvolution 85
Stack normalisation 62, 64
    effects of 64
Stack panels 68
Stacking charts 52, *53*
Stacking, common mid-point 61–4
Stacking velocity 64–5, 70, 71, 116
    low 92
    problem with 128
Stacking velocity analysis methods 66
Stacking velocity derivation 64–70
Stacking velocity function 66
Stacking velocity interpretation and checking routines 70
Standard tape formats 4, 47–50
Static correction 122
Statics, field 121
Stationary phase region 133, *135*
Statistical deconvolution, using Wiener filter 82–7
Statistical method, resolution improvement 76
Step model 141

Stoneley wave 147–9
Stoneley wave phase velocities 156
Summation, simple 66
Surface stacking charts 92
    and source array simulation *93*
Survey maps 115
Surveys 106
Sync code 48, 49

Tape controller 3
Tape drives and tapes 3–5
Tape reel header record, SEG Y 50
Tape, storage capacity of 4–5
Tape usage, efficiency of 4
Taper-off 52
Taper-on 52
Temporal resolution 72–87
Testing schemes (marine seismic data process) 110–16
Three (3)-D geometry and processing
        considerations 119–21
Time domain, migrated 125–7
Time series
    continuous and discrete 19
    definition 9
    seismological 9–10, 11
Time series analysis 9–46
Time shifts and extended arrays 93
Time shifts and the DFT 14
Time-variant filtering (TVF) 111, 114
Toeplitz structure 34, 151
Tomography 156–62
    algebraic reconstruction techniques (ART) 161–2
    convolutional methods (filtered back-projection) 159–61
    Fourier transform methods 158–9
    matrix inversion 157–8
Trace azimuth distribution 119, *120*
Trace balancing 105
    before display 103–4
Trace centroid 118–19, *120*
Trace invariant scaling curve 57, *58*
Trace offset distribution 119, *120*
Trace summing and mixing 86–7
Trace sums 52
Trade-off, resolution and parameter uncertainty *146*, 147
Transient sampling theorem 35
Transmission loss 56
Two (2)-D convolutional filtering 105
Two (2)-D dip filters 93
Two (2)-D filtering 39, 41, 94
Two (2)-D Fourier transform 38–46, 159

UK Offshore Operator's Association (UKOOA) 116
Uncertainty in parameter estimates 139
Uncertainty zones *132*
Unit-delay operator 22
Unmigrated time domain 127, *129*
Unwanted signal
    attenuation of 89–100
    recognition of 89

Variable area/wiggle trace trace (VAW) mode and
        display *100*, 101, *102*
Variable filters, low/high cut 114
Vector plotting 6, *7*
Velocities and velocity sensitivity in migration 128–32

Velocity analyses   97, *98, 99*
  CDP gathers   123
  horizon oriented   109
Velocity analysis   121, 122
Velocity analysis enhancement   68
Velocity analysis techniques   68
Velocity fields   1
Velocity interpretation and salt diapirs   109
Velocity, problems with   127
Velocity relationships between the 3 domains   129, *131*
Velocity spectrum   66
Velocity stacks, constant and function   122
Velocity, types of   64–5
Velocity uncertainty   132
Vertical (file) sums   52
Vertical seismic profiling (VSP)   152
Vibrational noise   55
Vibroseis   122
Virga
  of aliassed frequencies   137
  of dispersed frequencies   137
VSP   152
VSP processing   151–4

Wave equation
  differential   125
  Kirchhoff integral solution   124
  scalar   124–5, 133, 162
Wave separation, a linear inverse problem   153
Wave theory, simple   51
Wave-equation migration   124, 125
Wavelet compression   105, 111
Wavelet, criteria for selection   72
Wavelet distortion   71
Wavelet estimation   82
Wavelet length (effective)   70
Wavelet shapes   *74*, 75
Wavelets
  interference   75–6

  mixed phase   24
  zero-phase   23–4
Waves (up and downgoing), separation of   152–4
Weighted stack   96, *97*
Well-log information, use of   129
Well-log velocities   70
Whetstone rating   2
Whetstone suite   2
White light   34, 86, 111, 151
Whitening   72, 85
Whittaker–Nyquist–Shannon (WNS) criterion   13
Wiener filter   31–4, 76
  uses of   32–4
Wiener filter design
  and auto-correlation   27
  and white light   34
  practical   76–8
Wiener filter theory   28
Wiener inverse filter   23
Wiener inversion   33
Wiener–Hopf equation, discrete   32
Wiener–Levinson algorithm   85
Window carpentry   36
Windows (adjacent), merging of   86, *87*
Windows and data dependent gain routines   104
Word length   3
Word organisation   3
Wrap-round   18, 94
  and spatial aliassing   42

Zeroing *see* muting
Zero-offset seismic time section   123
Zero-phase filter   87
Zero-phase section, advantages of   74
Zero-phase wavelets   23–4, 74
  and minimum-time duration   23–5
Zone amplitude level   57
Z-transform   22, 23